BREEDING THE WORLD'S LARGEST LIVING ARACHNID

AMBLYPYGID BIOLOGY, NATURAL HISTORY, AND CAPTIVE HUSBANDRY

Orin McMonigle

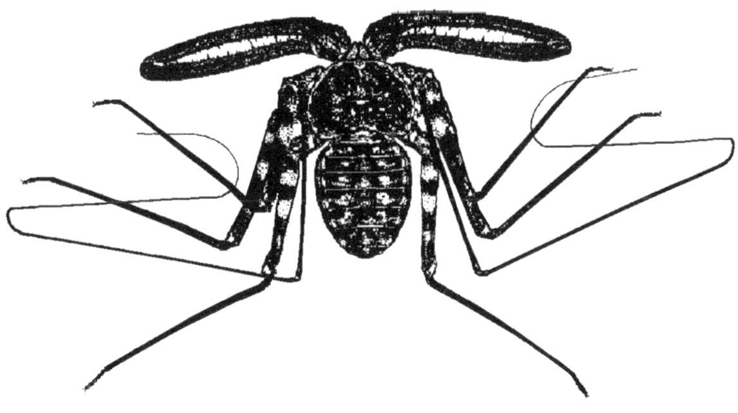

COACHWHIP PUBLICATIONS
GREENVILLE, OHIO

DEDICATION

To my father, whose aquariums proved fascinating laboratories once the offending water was removed.

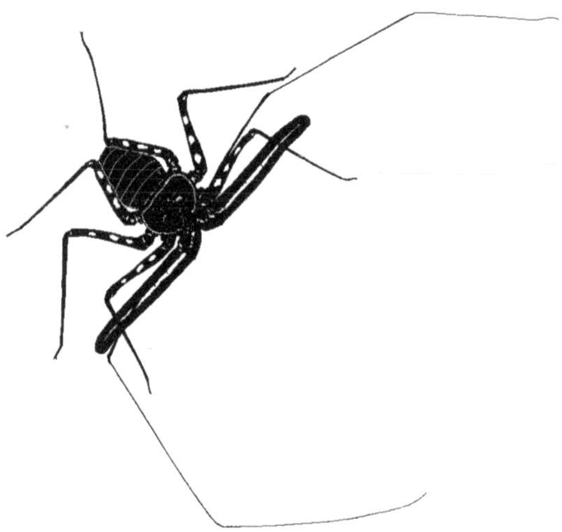

Breeding the World's Largest Living Arachnid

ISBN 1-61646-183-7
ISBN-13 978-1-61646-183-6
CoachwhipBooks.com

Front cover: *Heterophrynus batesii* and young.
Back cover: *Heterophrynus batesii* and young, Day 11 (starting to molt).

CONTENTS

Male and female *Paraphrynus mexicanus* can be difficult to sex dorsally if the female has not eaten much, but in this case it is easy (male at top).

ACKNOWLEDGEMENTS

I thank God for a world full of an endless parade of amazing creatures, and for my ability and desire to work with them. I thank Dr. Linda Rayor for providing editorial comments, husbandry experiences, and sociality charts, Christian Schwarz, Wade Harrell, Michael Seiter, Tammy Wolfe, Martin Hauser, Richard Trone, Christian Elowsky, and Sean Sutor for editorial suggestions. Great thanks is also extended to Matt Kaplinsky for the *Heterophrynus alces* specimens and personal experiences with *Phrynus barbadensis*, Sam Floyd for import details of *Heterophrynus batesii* and experiences with *Paraphrynus raptator*, Paul Bollinger for *Phrynus marginemaculatus* collection experiences, Eddy Hijmensen for habitat details relating to *Acanthophrynus coronatus*, Todd Gearheart for information on imports of *Heterophrynus alces* and *Phrynus barbadensis*, Connie Hurt's observation of *Heterophrynus elaphus* in Peru, Tripp Davenport for *Phrynus operculatus* habitat information, Brian Hodge for *Phrynichus exopthalmus* husbandry experiences, Christian Schwarz for habitat observations of *Damon medius* and *Charon grayi*, and the Cleveland Museum of Natural History for lending an out-of-print arachnid book with a detailed key to Amazonian amblypygids.

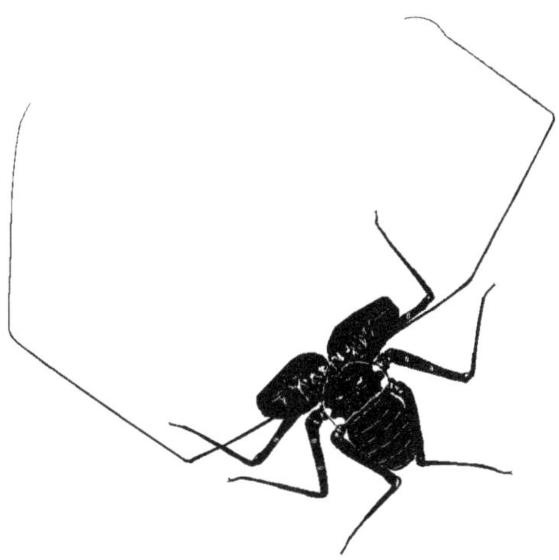

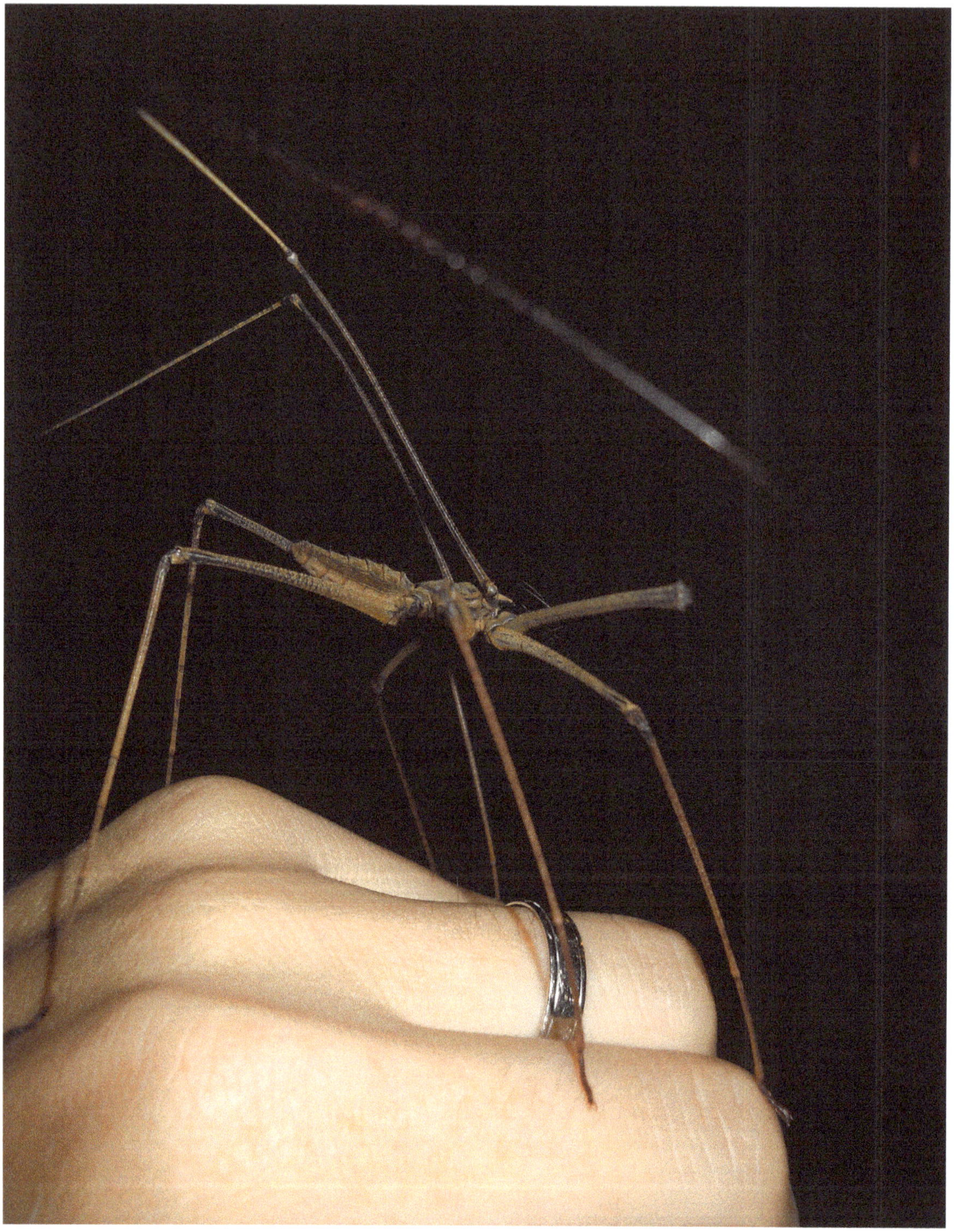

Whipspider dimensions.

INTRODUCTION

THE PURPOSE of this book is to share culturing experiences for these strange, oddly beautiful, and amazing arachnids. The primary goal is to provide readily accessible information on whipspider reproductive biology and repeatable husbandry parameters in a large format with photos that display the beauty of these amazing creatures. The whipspiders are by far my favorite arachnid since they are large, possess unusual beauty, are harmless to humans, do not immediately murder each other, and can be reared successfully in captivity through multiple generations. It is hoped the reproductive and husbandry reports shared here will encourage expanded and continued captive breeding efforts. Hopefully the incredibly eerie beauty displayed in the photographs piques the interest of those who are considering keeping these fantastic creatures as pets. Though less likely, it would be wonderful if increased awareness and appreciation of this order eventually reaches some of the local peoples who fear them. Countless harmless creatures are smashed every year by humans who fear whipspiders are dangerous.

Despite a frightening and intimidating appearance, whipspiders are gentle giants, completely harmless to humans, and make excellent pets. They take little effort and give so much back in observable and fascinating behavior. Whipspiders do not produce venom like spiders and scorpions and do not carry disease like ticks and mites. They produce no unpleasant defensive chemicals like vinegaroons and harvestmen. To avoid predation they rely on defensive abilities that cannot harm humans:

"Scary whipspider!"

speed, stealth, and rarely, intimidation. Large amblypygids, at the very worst, are only capable of inflicting superficial damage to a human's finger with their spines on the palps and chelicerae if grabbed roughly. The largest and most aggressive species kept so far are rarely capable of drawing even a tiny droplet of blood. In addition to a gentle demeanor, whipspiders make great pets due to their minimal requirements for housing and feeding. Inexpensive deli-cups, plastic storage boxes, gallon jars, and glass terraria make great habitats and take up little room. There are no special filters, expensive lights, or mechanical devices required, often just a large piece of bark. In captivity specimens will feed on crickets, which are low cost and readily accessible. Other commercially available or easily cultured food items are optional. (A short section on feeders is included since prey items can be cultured easily and cheaply.) As cold-blooded predators the majority do not require constant feeding and they can be very healthy if only checked on every week or two. If I were forced to leave town for a month, I could feed and water them before leaving and reasonably expect every animal to be alive and healthy on my return.

The unique behavior of whipspiders is endearing to the invertebrate enthusiast, whether solitary or in community. The slow, deliberate movements during hunting, courtship, or exploration of the habitat are fascinating to observe. Whipspiders carefully unfold improbably long, antenniform front legs to probe their surroundings and search around corners and inside holes. These rapidly spring back and fold in at the slightest disturbance to allow animals to run sideways at breakneck speed. Interactions between siblings and cage mates are usually gentle and their sociality appeals to the casual and scientific observer.

The Amblypygi have long been a neglected group in the literature. A number of general arachnid books offer some information on this group but it is usually extremely limited, lacking in the most basic of husbandry details, and second-hand. *Whip Spiders (Chelicerata: Amblypygi) Their Biology, Morphology and Systematics* (2000) by Weygoldt is an amazing book the whipspider enthusiast should not do without, but it is a technical text on biology which offers just a few pages in the way of captive husbandry details. It is in black and white and primarily focuses on Old World species and species that have not

"Oh, it's harmless."

made it to the North American enthusiast. *Whipscorpions and Whipspiders: Culturing Gentle Monsters* (2008) by myself is the only work I know of offering repeatable, first-hand husbandry advice of any depth, but it is only forty pages, has just a few color plates and includes whipscorpions.

There have been some articles published in a few hobbyist magazines over the years pertaining to amblypygid care including limited reports on various social behaviors, but articles have been uncommon and obscure. The few, rare husbandry articles are possibly all cited in the bibliography of this text. A number of Weygoldt's works reference or, more often, provide short comments in reference to lab rearing. There have also been a variety of taxonomic, systematic, behavioral, and ecological articles in various journals. Nearly all these short resources are hard to find, and combined would not provide much useful information for those interested in keeping thriving amblypygid cultures. A pair of scientific papers published (Rayor and Taylor 2006, Walsh and Rayor 2008) focused on experiments in whipspider social behavior with possible application to husbandry. These offered new insight into social behavior that was not discussed in previous works, but covered only limited aspects of two species. There is still very much to be learned and understood in whipspider social interaction.

Other than the paltry literature outlined, there is little available information on care and culture for these amazing arachnids available in print. What is in print is composed of difficult-to-acquire scientific papers, non-English articles, or unbelievably brief or inaccurate, shallow descriptions in general arachnid or small

From the cover of *Culturing Gentle Monsters.*

animal guides. For this reason, the text you are holding brings together the amblypygid portion of *Culturing Gentle Monsters*, half a dozen articles on amblypygids from *Invertebrates-Magazine*, as well as many new reports, experiments, and photographs that have never been published before. Although they have been hatched repeatedly in captivity and are very fascinating and beautiful in similar ways, whipscorpions (the other half of *Gentle Monsters*) are not included here as much because they vary from whipspiders as because there has not been and may never be a viable opportunity for them to be continuously cultured in captivity. Beyond the extensive updates, I am excited to be able to include new species in cultivation as well

Mother with 1st instars.

as chapters on health and sociality. In my 2008 book I was forced to leave out a lot of information due to the inherent limitations of the format. Since then I have also gained experiences with additional genera and species, including the victory of rearing and breeding the "Holy Grail" of amblypygids, *Heterophrynus batesii*. This creature has unbelievable palps, a world record whipspan, and thanks to the 1997 *Fauna* magazine article/pictorial, it is squarely placed atop the wish list of every US whipspider enthusiast who saw it. Thanks to Coachwhip, this book will provide a large format, hardcover, color text deserving of the Amblypygi.

Since representatives of this handsome and harmless arachnid order can be legally imported it is surprising how limited our selection in the United States has been. Until recently the captive amblypygid hobby was unbearably sparse, with only two species seen by any but the quickest and luckiest. For many years US hobbyists have been essentially limited to our native *Phrynus marginemaculatus* from Florida and African imports of *Damon diadema* from Tanzania. More than a decade ago there was nothing a normal hobbyist could find other than *P. marginemaculatus*, and even this common species took some effort and time to locate. Specimens have been generally available as wild-caught, but almost never as captive-hatched since they start out too small to generate interest. Acquiring the large African *Damon* pre-2001 was nearly impossible, but since 2001 it has been available most of the time. The

Tanzanian whipspider has been available as wild-caught adults, captive-hatched immatures, and even captive-bred young. In 2007 and 2008 a very limited number of specimens of *Damon medius*, a creature which looks like a bulkier, more gray *D. diadema*, showed up a few times. The enigmatic *Paraphrynus* from Arizona, supposedly looking much like *P. marginemaculatus,* has shown up a few times over the years, though not on any vendor lists. Another half-dozen species have shown up to some degree in the last five years since *Culturing Gentle Monsters*. In 2009 *Heterophrynus alces* arrived and in 2010 *Heterophrynus batesii* finally made it to a few vendors' lists. Around the same time some *Phrynichus exopthalmus* snuck in labeled as "Damon variegatus." The last few years in the United States have also seen other exciting new species of amblypygid make an appearance in the hobby, including *Phrynus barbadensis* and very limited numbers of our two *Paraphrynus*. This book is written to introduce these handsome species through images and to offer all currently available husbandry information.

There are a few possible explanations for our lack of choices, though none may be true since I remember seeing a few European whipspider lists with over a dozen species years ago. The first reason is the many members of this order generally appear similar and are colored only from brown to black. Even die-hard enthusiasts are not going to become greatly thrilled with acquiring a species that is remarkably similar to captive stock they already have. (I greatly regret passing up buying captive-hatched *Damon medius*, but at the time I had too many *D. diadema*, which do not look terribly different.) Nevertheless, there are some exotically different pedipalp armatures and varying leg attitudes and body ratios that differentiate certain species or groups. A second reason is some amblypygid species can be very sensitive to shipping, so import and collection stresses can be too much for them. Third, demand is not excessive but it stills seems greater than supply.

H. batesii male whips are longer than female's.

Despite recent fame, even the borrowed common name for this order suggests little appreciation of its standing as a unique arachnid order, as different from a scorpion as to a tick. The whipspider may have enjoyed some recognition in recent years, but not under the right name. *Damon diadema* touted as "cave spiders" were eaten alive by contestants on a popular network show in hopes of winning prizes. (There are, in fact, cave spiders, some of which are without eyes, but they are not amblypygids.) The most spectacular screen version must certainly be the oversized CGI creature copied almost perfectly from a living *Damon diadema* male that showed up on *Harry Potter and the Goblet of Fire* (2005)—the movie, not the book. It was referred to as a "harmless" creature, "lethal when she bites," and after a few minutes of wandering the classroom "she" was tortured and then put to death. In both reality and animation, whipspiders end up dead without a tear shed. Three-part eyes of the aliens in the 1988 *War of the Worlds* television series were the first thing I thought of on initially seeing a whipspider in the early 1990s. Similarities only begin with tetra eyes. Most renditions of *War of the Worlds* (1953) aliens display a large flattened central body holding the tetra eye mound (not eyes on stalks), long segmented tentacles like antenniform legs, and ambulatory appendages in multiples of three like the six whipspider legs and associated triangular trichobothrial formations. Furthermore, nobody cries when countless numbers are slaughtered, and amblypygids are not credited even in passing.

Exuvium. *Heterophrynus batesii* 4th instar.

The two most commonly used names for the Amblypygi have been *whipspider* and *tailless whipscorpion*. The term whipspider is used preferentially in this book because tailless whipscorpion is too often confused with whipscorpion—even long-time enthusiasts often forget to add *tailless* to the beginning—and it is overly long and awkward. One might as well call spiders whipless tailless scorpions, though Evans (2007) offers a new combination in "tailless whip spiders." Another problem is, according to entomological common name rules, it would have to be spelled taillesswhipscorpion with no space, to show it is not a whipscorpion, an entirely different arachnid order. In addition to awkwardness and confusion, it is a borrowed name off a borrowed name and the only descriptive verbiage suggests it lacks something it possesses. Amblypygids possess small tail segments that play an important role in molting (McMonigle 2013b) but actually lack a flagellum, so if anything they should be called flagellumless-whipscorpions. When I first heard the term whipspider I thought it was a modern attempt to right the great injustice that is tailless whipscorpion. However, it is not a new name; the oldest reference I have come across listing a common name for the Amblypygi uses whipspider (Stanek 1969).

Broodsac being extruded, "Is that a tail?"

Many long-time keepers unhappy with the horrendous tailless whipscorpion name prefer to use *amblypygid* since these creatures are one-of-a-kind arachnids. They are rather different from spiders but common names such as "whipling," "spiny palpster," or "amblypygid," with absolutely no ability to convey meaning to the average person, are not really common names. When I display invertebrates at an event and someone asks what *that* animal is, the answer amblypygid is usually promptly followed by the obvious response: "What is that?" or simply a disapproving or annoyed look. The answer taillesswhipscorpion is better accepted, but taken as nearly as nonsensical as it is. Whipspider may not be perfect but it proves reasonable and functional. It is also more systematically accurate.

Whipspider not whip spider. A plant bug is a bug but a ladybug is a beetle. In the same way a butterfly and dragonfly are not flies, while a house fly and horse fly are; the list goes on. A whipspider with a space in middle would be a type of spider, not an amblypygid. Some texts use a hyphen in the middle to differentiate the name from the true spiders, but a hyphen conveys no added meaning and for this reason hyphens are purposefully excluded from most common name schemes except where the meaning would be lost (for example, w-marked cutworm). Rules for invertebrate common name use of contractions and hyphens are not new but were developed over a hundred years ago by the Entomological Society of America and have been adopted by the American Arachnological Society, the American Tarantula Society, and others (Breene 1995). The problem with a space in the middle is it tells people these animals are spiders and members of the Order Araneae, not Amblypygi.

Bias against the Amblypygi even among arachnologists is hard to explain, but may relate to the fact many are only familiar

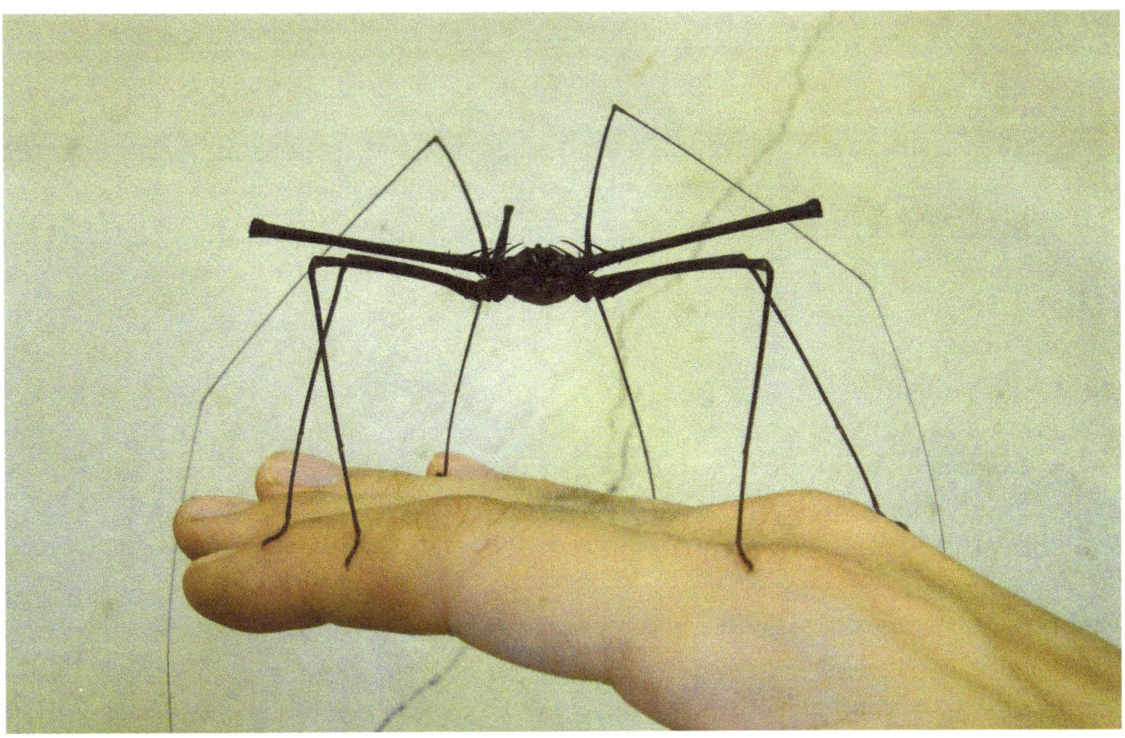

Holy Grail.

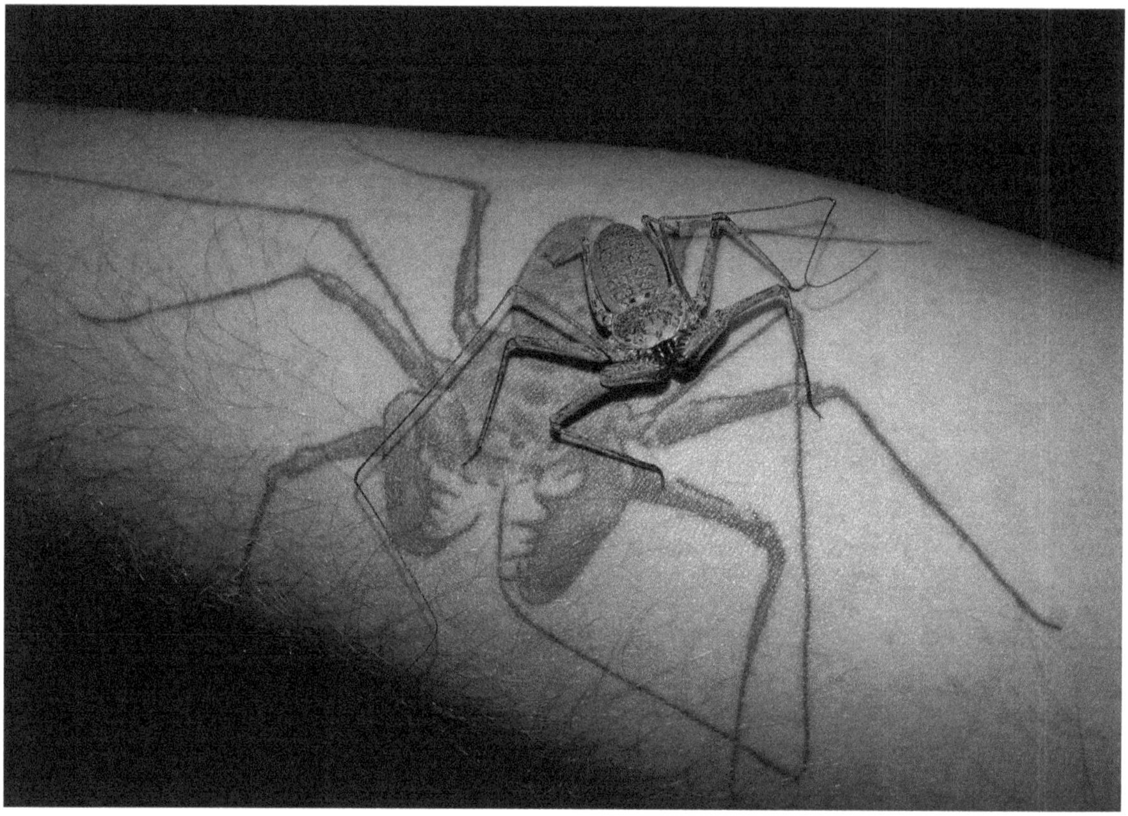

The only thing more difficult to find in the US than this *Paraphrynus raptator* is an amblypygid tattoo. © Paul Bollinger

with this order from pictures. American and European arachnologists almost never see these in the field and are not worried about what to call them. Piles of books that discuss arachnids do not even mention this order (and a few other uncommonly seen orders) and those that do rarely give it a paragraph. The refusal to accept a less confusing common name by one or more US entomological organizations is not entirely understood, but may simply be a lack of desire to address a common name problem for a rarely considered order. (A common name at its most basic must be commonly used.) The only unbelievable oversight relates to discussion of the maximum dimensions of arachnids. The largest whipspider possesses double the legspan of the next largest extant arachnid, a spider. Yet, it is perfectly acceptable for book authors, for example Becalloni (2009), to unashamedly list the maximum legspan of a spider at 11.8″ (300 mm) immediately above a listing for a body length of 1.8″ (45 mm) for a whipspider—both under the heading "Size"—while wholly ignoring the latter's undisputed 23.6″ (600 mm) documented legspan (Weygoldt 2000). The odd, skinny tail-abdomen of the scorpion explains its departure from direct comparisons, but the indiscriminate flip-flopping of legspan and body length for other orders in comparisons with spiders is unjustified. Even if an honest case could be made to reject all but ambulatory legs, the whipspider maximum walking-leg span would easily give the largest documented spider a run for its money.

None of the *Damon diadema* specimens in the photographs are wild caught; most are second to fourth generation, while the *Heterophrynus batesii* are F1 and F2 captive generations from wild-caught parents (P) with the exception of one wild male. The initial adults of both species were acquired as 2nd instar hatched in captivity (F1) and reared to maturity in captivity. With exception of the captive-hatched immatures, the specimens depicted for other species were originally wild caught. Background in photographs not expressly labeled natural habitat should be considered representative of lab materials, not natural surroundings.

In this text I offer husbandry advice uncovered over two decades of keeping and breeding whipspiders. There are certainly those who have seen and kept more wild specimens, or are far more versed in taxonomy, but few, if anyone, would claim to have the same level of experience rearing whipspiders through multiple consecutive generations. Take this into account when weighing the options and husbandry advice contained in the text. The advice given here will make all the difference to new keepers and may even make a difference to experienced keepers. There are many discoveries even beginning hobbyists can make since little to nothing has been documented for many species. I learn new things about these amazing creatures all the time. I have been thrilled by successes and depressed by failures. Failure is a wonderful learning tool as long as the cause can be determined and the cause is not something foolishly repeated like a freshly molted animal jumping to a hard floor during handling, forgetting to water, or not seeing a vicious, hiding cricket. Since it is difficult to obtain livestock I hope that by sharing my experiences, others can use my failures and successes to establish breeding lines of species that disappear as quickly as they show up.

The best efforts and hours, if not years, of research have gone into finding the most recent classification information for the whipspiders in this text. However, taxonomy is difficult for many of these creatures since external characteristics can be variable and difficult to identify. Species found in the hobby can be easily identified, but new, unidentified species may enter now and then. It can be difficult to identify some species since not all the original descriptions offer characters that can be useful in identification (Weygoldt 2002). However, many species are identifiable and in recent years much of the identification details have become accessible to the inquisitive enthusiast.

These handsome arachnids are truly amazing animals and superior pets. *Breeding the World's Largest Living Arachnid: Amblypygid Biology, Reproduction, Natural History, and Captive Husbandry* provides detailed information for the successful husbandry of these creatures. Captive breeding is personally rewarding and is important for both the future of the hobby and the whipspiders. With man's unstoppable habitat destruction around the world, captivity is likely one day to be the only possible refuge for some species or they will be lost entirely. Humans do not always save habitat for creatures they love and know. Hopefully this text will increase both for the whipspiders.

1
WHIPSPIDERS AMONG THE ARACHNIDS

Class Arachnida
- Order Amblypygi—Whipspiders
- Order Araneae—Spiders
- Order Thelyphonida (Uropygi)—Whipscorpions
- Order Schizomida—Short-tailed Whipscorpions
- Order Palpigrades—Microwhipscorpions
- Order Opiliones—Harvestmen
- Order Scorpiones—Scorpions
- Order Pseudoscorpiones—Pseudoscorpions
- Order Solpugida (Solifugae)—Windscorpions
- Order Ricinulei—Hooded Tickspiders
- Order Acari—Mites and Ticks

There are eleven extant orders of Arachnida including the familiar spiders, scorpions, mites and ticks, and harvestmen. Four of the arachnid orders include the term *whip* in their common name because the front legs are modified into elongate tactile organs that function like the antennae of other arthropods and are not functional for locomotion. However, many members of most orders use one or more pairs of legs, which are visibly modified to varying degrees, to detect surroundings and movement. In the Opiliones, Ricinulei, Solpugida, and others, the sensory legs tend to be visibly modified but still commonly touch the ground during locomotion and are not considered antenniform. A brief explanation, differences, and similarities the various orders share with amblypygids are outlined in this chapter.

Previously the Amblypygi, Thelyphonida, and Schizomida were considered part of a single order, the Pedipalpi, but recent molecular studies have found the amblypygi are more closely related to the Araneae and so are the sister taxon to spiders (Weygoldt 2000). Reproduction and early development is similar for the orders formerly comprising the Pedipalpi, though there is some correlation to development of Pseudoscorpiones and Scorpiones. Maternal care is extremely important and is required until immatures reach 2nd instar. (Maternal care may continue on much longer, but afterwards is no longer life or death.) Eggs develop inside the female's abdomen for a few months and then are extruded inside a membranous, fluid-filled sac (the eggs are never exposed to air). Inside this broodsac they are nourished, or at the least hydrated, by fluids from the female's oviduct until the postembryos are ready to hatch into 1st instars. Newly hatched immatures, 1st instars, cling to the female's body until the

Whipspider.

first molt. They tend to encircle the opisthosoma (different from scorpion 1st instars that stay only on top of the mother's opisthosoma). At this stage they have limited mobility, underdeveloped appendages, lack protective coloration, and do not feed. Once they molt into 2nd instar they look like miniature adults, leave the mother's opisthosoma, and are ready to hunt for food. Immatures will eat live prey and molt a set number of times (commonly four to six) before reaching sexual maturity. After reaching maturity they live multiple years and females tend to produce a single, often seasonal brood, yearly. Adult whipscorpions and short-tailed whipscorpions do not molt after adulthood, but whipspiders continue to molt.

Damon diadema molt to 2nd instar.

Whipscorpion palps terminate in an opposable tarsal claw formed by the two tarsal segments. This pincer is used for burrowing and prey capture.

Whipspiders have a compact, round body shape which superficially resembles a spider's but their armored, segmented abdomen is very different from the membranous abdomen of all but the most primitive Liphistiomorph spiders, which also have a segmented abdomen. These primitive spiders are rare, representing less than 1% of spider species. Although whipspiders have the characteristic eight legs of arachnids, there are only six walking legs since the front legs are useless for locomotion. Whipspiders never make silk. They have eight eyes; two are located at the middle front of the prosoma and a cluster of three eyes is found on each side. The eye clusters on the sides are closer to the front and more noticeable than on whipscorpions. Species that live deep inside caves or in ant nests can have more or less reduced eyes. The extremely elongate antenniform front legs are highly segmented, particularly in the tibia and metatarsal segments, and much longer than the back three pair of legs. This legspan (or whipspan) is rarely less than a few inches and large male specimens of at least three species may exceed two feet (>610 mm) in span. The spine-covered, raptorial pair of prominent, folded mouthparts (palps, short for pedipalps) are normally much more massive than the walking legs and are important for feeding, mating, and fighting. These grabbing mouthparts function like front arms and look like an oversized pair of arms when in use. Palps do not end in an opposable tarsal pincer like those of the scorpions, whipscorpions, and pseudoscorpions (the tarsus folds up against the tibia and usually cannot completely close), but in a single, sharp, stiletto-like spine. They have two pairs of book lungs under the opisthosoma like whipscorpions, scorpions, and mygalomorph spiders. Amblypygids have a small "tail" composed of the reduced opisthosomal segments, but it never ends in a flagellum like the whipscorpions, and the segments are half ring rather than circular. This order has proven somewhat easy to rear in captivity through multiple consecutive generations (McMonigle 2008), while captive-hatched and captive-born are common for the few available species. There are presently 161 described species in five families and seventeen genera (Prendini 2011c).

In contrast, while spiders also usually have eight eyes, they are arranged differently in a double row at the front of the prosoma. The number of eyes can vary; some spider groups have six eyes or less and there are even eyeless cave spiders. Spiders use all eight legs for walking and do not have anything like the antenniform legs of the amblypygids. Although spider legs are covered in sensory hairs, none of their legs have the extensive joints that allow amblypygids to move their whips 360° around their bodies. Spider palps are also very different. They often look very much like an extra set of legs, which is the reason this arachnid mouthpart is most commonly known as a pedipalp ("pedi-" meaning foot) for all arachnids. However in some spiders and the majority of other arachnids, the "pedi"-palp rarely looks anything like a leg. Like whipspiders, the chelicerae are composed of a large basal segment and end in a fang, not a pincer. Also, they have a narrow waist between the prosoma and the rounded abdomen. Spiders produce silk from the end of the abdomen for various purposes, from catching prey to sperm transfer and egg protection. The male creates a sperm web and uses structures on his pedipalps to store and transfer sperm to the female. The female wraps large clusters of eggs, often in the hundreds, in a silk capsule commonly referred to as an eggsac. Eggs are exposed to the air before deposition in the eggsac, there is no fluid in the sac, and so they can receive no nourishment or moisture from the mother. Females of seasonal species can often produce three to eight eggsacs in just a few months. The eggsac is not always abandoned but may be guarded or carried around by the mother. Rarely, as in wolf spiders, mothers also carry around early instars on the opisthosoma. For the vast majority of spider species the adults are short-lived, seasonal, and do not molt after maturity (female tarantulas being among the few exceptions). While amblypygids possess no venom, nearly all spiders have venom glands in the chelicerae and a few

Araneae: Pedipalps, book lungs,
and the male's sperm web are visible.

Whipscorpions have far less complex antenniform forelegs with very long patellae. *Mastigoproctus giganteus* are highly aggressive towards each other and these two had to be manually separated to prevent death and dismemberment.

are potentially dangerous to humans. Spiders are the most familiar arachnid group to humans, and the spiders in the Family Theraphosidae (popularly known as "tarantulas") are the most popular group kept and bred by enthusiasts, with hundreds of species in culture. Spiders used to comprise the largest known order with around 43,000 described species, but have been surpassed by the Acari (mites and ticks). With so many species, they are highly variable in size, and adults can range from barely visible to somewhere less than a foot (<300 mm) across the legs at the furthest point for the largest specimen of the largest species.

The uropygids or whipscorpions were recently renamed Thelyphonida because the Order Uropygi originally included Schizomida. However, uropygid is still an accepted common name. Whipscorpions are commonly known as vinegaroons in the United States and vinagrillos in Mexico due to the acetic acid component of the defensive spray. (The fluid smells strongly like vinegar, which likewise contains dilute acetic acid.) Vinegar may be used to flavor human food but it must be repulsive to some predator. Vinegaroon is a very appropriate common name since members of this order are very different from scorpions and uropygid has little meaning, especially now that the order name is no longer valid. The vinegaroon body is similar to the whipspider but much more elongate. It superficially looks more like a scorpion, while the pedipalps terminate in small, opposable tarsal segments that, except for an extra tibial spine, look like the scorpion pincer. The chelicerae, however, are two-segmented like spiders and do not end in a claw like the three-segmented chelicerae of scorpions. The antenniform legs have extremely long patellae, four to five times the walking leg patella length. They are not segmented at the tibia, nor as highly developed as whipspiders, but similarly end in vestigial, microscopic tarsal "claws." The eye placement seems similar to amblypygids but is different. The large median pair is located at the front of the prosoma against a raised ridge, somewhat like scorpions, but not located on a defined plate (or ocularium) like amblypygids and most Opiliones. The two lateral clusters of three

Mastigoproctus giganteus broodsacs are similar to whipspiders, internal fluid level is greatest immediately following initial formation as seen here. Over the next few days excess fluid is withdrawn back into the female's body.

eyes are located further back to the sides. (In scorpions the lateral eyes are forward of the median eye mound.) The opisthosoma contains book lungs and is connected by a narrow waist to the prosoma like amblypygids and spiders. It ends in three very small circular ring segments and terminates in an incredibly long, whip-like flagellum. The tiny tail-like opisthosomal segments add almost nothing to body length, unlike the true scorpion's tail-like continuation of the abdominal segments. (Whipscorpions have no metasoma nor telson.) Glands in the opisthosoma that open near the base of the flagellum produce and secrete the defensive fluid and the flagellum helps to fling and direct this fluid towards the threat. These are popular in captivity because they are gigantic and scary looking, but are harmless, handsome, handleable, and hardy captives. Whipscorpions have been hatched in captivity but after more than twenty years of somewhat regular husbandry there has not yet been documented F2. This is partly because it takes at least four years to rear *Mastigoproctus giganteus* to maturity, but primarily because it is not nearly so easy to produce young as it is to mate or keep alive. Gestation is slow and often unsuccessful, but results in 30-40 large, hardy immatures. Gestational difficulties may have a dietary component, since *M. giganteus* may feed more on scorpions in nature, while specimens are fed only crickets and cockroaches in captivity. There are 110 species (Prendini 2011d) worldwide with a majority of described species from Southeast Asia. Although there are not terribly fewer species than whipspiders, there is only one family and very little deviation in form. The largest species ever kept in captivity, and arguably the largest in the world, is a common USA native, our only native species, *M. giganteus*. Large specimens are up to 3″ (76 mm) in body length and nearly double that when the flagellum is included. Inclusion of the flagellum seems to be what leads to tall tales (not tails) of "unnamed exotics" reaching 6″

(152 mm). Fortunately the myth does not double again in size every so many years.

Schizomida are neat little creatures that rarely exceed 1/3″ (8 mm) in length. They look very much like miniaturized whipscorpions except the flagellum is short and the male's opithosoma appears to end in a small globular structure. Orientation of the palps is also different, as they bend vertically rather than horizontally. Reproduction is similar to whipscorpions but with far fewer eggs. Less than ten young hatch after about three weeks (Sejna 2005). They also experience five postembryonal instars. The adults do not molt and can survive a few years in captivity. Nearly all species are found in tropical areas and are uncommon. According to Levi and Levi (1968) at least one species is found in southern California, *Hubbardia pentapeltis*. There is also a tiny species in Florida, *Stenochrus portoricensis*, but I have never heard of captive culturing or seen either for sale or trade. There are two families and 260 species worldwide (Prendini 2011a).

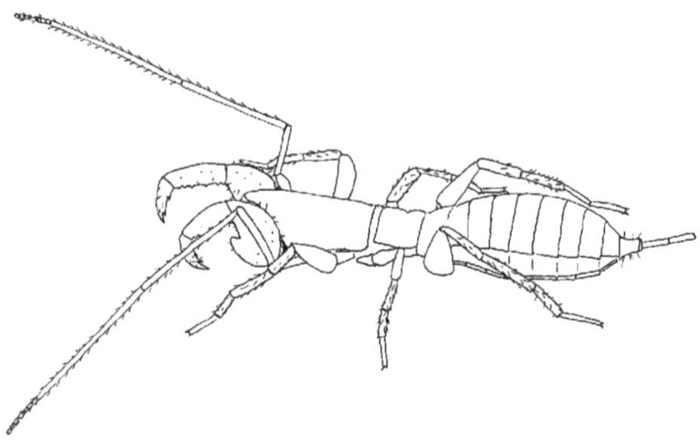

Schizomida female.

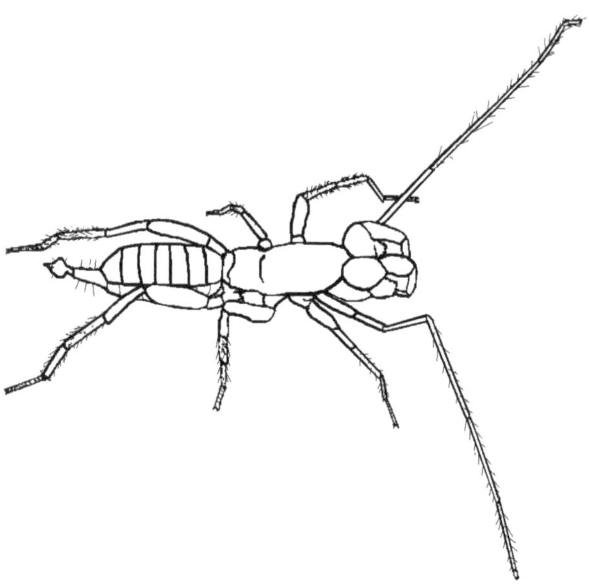

Schizomida male.

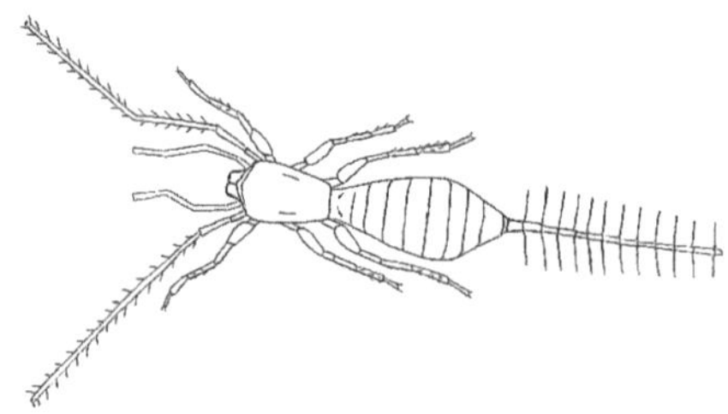

Palpigrade.

Palpigrades at less than a tenth of an inch (2 mm) make the Schizomida seem huge. They are subterranean and are generally rare. Specimens have not been seen in the hobby and the means of reproduction undocumented. They are very different and the pedipalps look more like walking legs than the crushing, powerful weapons of the other "whipscorpion" orders. They are mostly tropical like all the small orders, but a few species are found in the southwestern US. Being difficult to find and restricted in range, their near-microscopic size is an added detraction to most enthusiasts. There are around eighty known species (Harvey 2003).

Opiliones superficially resemble small amblypygids but segments of the prosoma and opisthosoma are fused so they only have one body segment, while the chelicerae end in claws like scorpions and most

Captive-reared adults bred from Florida, USA, stock, *Vonones ornata*, Order Opiliones, Suborder Laniatores, Family Cosmetidae.

Acari. The median eye mound is similarly located but there are no lateral eye mounds. The second rather than first pair of legs is elongate and used like antennae, but these appendages do not look terribly different from the walking legs and they lack advanced articulation. Many members of the suborder Laniatores have short legs held off to the sides, superficially resembling *Phrynus* as they also squeeze under bark and into narrow crevices. Laniatores include the largest and most colorful tropical species, a few with long legs, but the species with the longest body is a palpatore in the Family Trogulidae. This species, *Trogulus torosus,* can measure 7/8″ (22 mm). It looks like a gray pumpkin seed with a legspan less than triple the length of the body. The most familiar harvestmen are the long-legged members of the suborder Palpatores with legspans that can reach 5.5″ (14 cm). The long stilt-like legs are highly suggestive of *Heterophrynus* or *Paraphrynus* whipspiders. However, the legs of both suborders are only superficially similar and none of the Opiliones can move with the incredible speed of the Amblypygi. A number of harvestmen, primarily long-legged species of the suborder Palpatores, have enlarged, spine-covered pedipalps that resemble those of whipspiders. The elongate palps are sexually dimorphic and I have seen a few species in the US with folded palps longer than the body. Some male Palpatores, including *Megalopsalis* spp. from New Zealand (Forster and

Forster 1970), have folded palps greater than three times the width of the body like the most exaggerated whipspiders, but these are held forward rather than off to the sides and the animals themselves are relatively tiny. They are likewise harmless to humans and have no venom, but have scent glands on the prosoma that produce various chemicals including quinones and phenols whose primary purpose is thought to be defense (Kury and Rocha 2002). Reproduction is entirely different, as males have a penis and employ direct fertilization while females have an ovipositor and bury eggs. Adults are often communal but they tend to eat immatures in captivity (McMonigle 2012b). The more extravagant tropical Opiliones could make fascinating pets but the largest are smaller than most small *Phrynus* species and they have been, at best, very rarely available. There are more than 6,000 species.

Most of the "tail" is part of the abdomen as the anus is located just before the telson. In the longest species these segments can be incredibly long and thin, *Centruroides* bark scorpion, Arizona.

Uroctonus mordax mother with 1st instar scorplings from northwestern USA.

Scorpions are outfitted with palps that terminate in gigantic tarsal pincers, three-part chelicerae each ending in a claw, four walking leg pairs, and a remarkably specialized opisthosoma ending in the telson, which is used to sting prey and predators. The unique opisthosoma is divided into the wide mesosoma, which contains book lungs and reproductive organs, and the tail-like metasoma that ends in a stinging telson. The telson includes the stinging apparatus of vesicle (venom bulb) and aculeus (stinger) and is unique to scorpions. It is not part of the abdomen, which ends with the fifth metasomal segment. Tails are often ignored in the body measurement of arthropods but no matter how thin and appendage-like the scorpion tail (metasoma) is, it is part of the abdomen in a way something like the flagellum of whipscorpions is not. Inclusion of the elongate, skinny metasoma of male *Hadogenes troglodytes* can give the longest specimens a 9″ body length (legspans are shorter). A less familiar and unique trait of scorpions is possession of trademark appendages, called pectines, located behind the rear legs. The comb-like pectines are organs used to detect vibrations, surface texture, moisture, odors, and possibly pheromones. The median eye pair is not located on a distinct plate while the lateral eye clusters vary in number and arrangement

Scorpion 2nd instars commonly remain on the mother's back after the first molt and ride around for a few weeks or more. Specimen from southwestern USA.

Amblypygi *Phrynus marginemaculatus* 1st instars usually attach to the bottom and top of the opisthosoma, this brood is a little irregular.

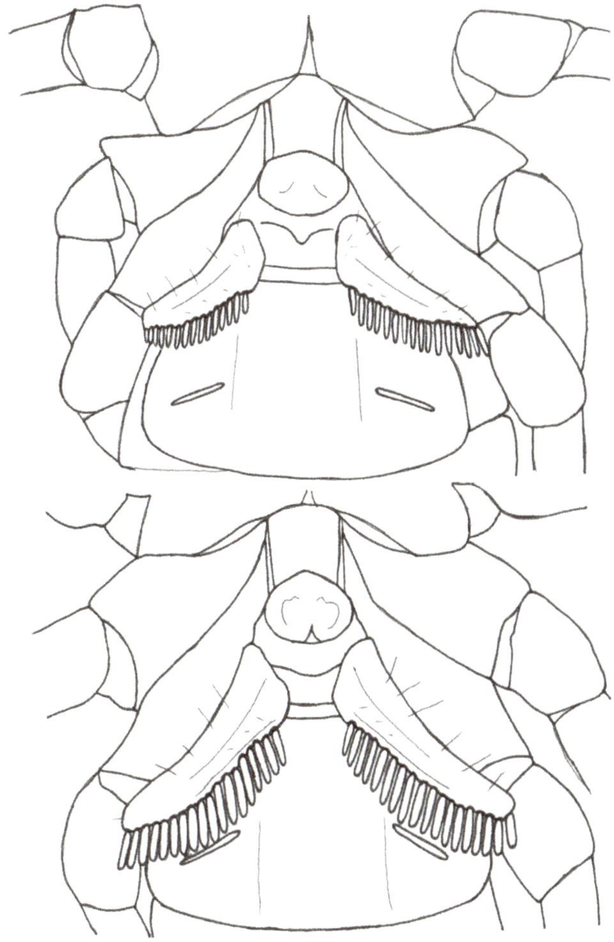

Scorpion pectines, *Pandinus imperator*, female top, male bottom.

(zero to five each). Males produce spermatophores taken up by the female during a courtship dance where the pair holds "hands" (pedipalp pincers). Females give birth to postembryos that immediately hatch into 1st instar immatures. They produce 10-100 eggs that are retained in the mesosoma and nourished internally until birth. As with amblypygids, a lot of time and energy is invested by the female in a single brood, eggs are never exposed to air, and time from mating to 1st instar is rarely less than six months. 1st instars are pale, do not feed, and cling to the mother's back until the first molt like whipspiders, but 2nd instars commonly remain on her back a few weeks or more following the molt (McMonigle 2009a). Of the 1500 or so species, only about two dozen have venom that is lethal to humans. However, some humans can become allergic to and die from any scorpion

Pseudoscorpion adult pair, *Microchernes dentatus*.

Pseudoscorpion broodsac with developing eggs. Eggs deform the surface of the broodsac but are not exposed to the air.

Pseudoscorpion postembryos enlarge greatly from nutrients provided by the mother through the genital operculum attached to the broodsac (compare the relative size of developing young with the abdominal volume to photo above).

venom. Otherwise, most are relatively harmless and second only to spiders as a favored pet.

Pseudoscorpions, as the name implies, are superficially similar to scorpions from the prosoma forwards, but the segmented abdomen does not end in a metasoma, there is no telson, chelicerae are two-segmented, and the venom is located in the claws of the pedipalps. These tiny creatures breathe through trachea and are more like windscorpions than scorpions. Like spiders they can spin silk, but with the chelicerae rather than abdominal appendages. (However, the male of one species uses an abdominal silk gland to form a web to guide the female to the spermatophore (Weygoldt 1969)). This order is more widespread than anything but Acari and Aranaea. Specimens are commonly found indoors and outdoors in cold temperate areas, as well as arid areas and the tropics. Captive propagation can be simple and undemanding because they survive in colonies, but are rarely kept because they are difficult to see. The largest species in the world does not reach 1/2" (12 mm) and legspans are

Pseudoscorpions form web chambers for molting, overwintering, and reproduction.

This *Eremobates* just laid a pile of eggs from only the right side of her body (usually eggs from both sides are laid together). Adult female windscorpions continue to feed and can lay three to five consecutive batches of eggs over a period of a few months before dying from old age.

rarely half as much more. Representatives of the same species live in close proximity and are tolerant of conspecifics, but will rapidly hunt down and attack other species in the same container (McMonigle 2009a). Males convince females to take up the spermatophore through a courtship dance like scorpions, complex webs, or unknown means. Females produce up to a few dozen eggs that hang from the abdomen in a brood pouch similar to whipspiders, but postembryos are nourished extensively inside the broodsac, and 1st instars do not cling to the mother's back. Females only produce young once or twice a year. There are twenty-six families with more than 3,454 species (Harvey 2011). Although most, if not all, are venomous, they are too small to puncture the surface of the skin and cannot deliver a noticeable bite to humans.

Windscorpions (Solpugida or Solifugae) are easy to differentiate from all other arachnids because of unique pedipalps and massive chelicerae, commonly larger than the head. Also, the cephalothorax has distinct head and thoracic regions, so together with the abdomen there are three visible body regions, while other arachnids have only one or two distinct body regions. These creatures are outfitted with long, notable setae and baton-shaped palps that end in a sticky, retractable organ which aids in prey capture and allows some species to climb

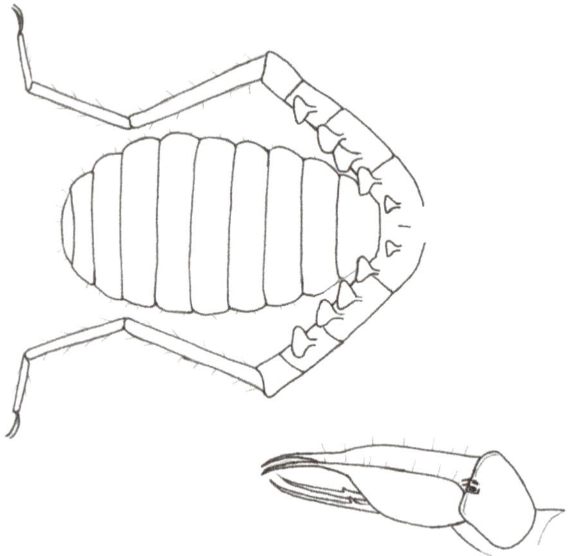

Windscorpion racquet organs
and cheliceral development.

smooth glass. (Such climbing looks like swinging from monkey bars and can be shockingly rapid in function.) Despite some misleading photographs and a bad B movie (*Camel Spiders* 2011), these arachnids do not make webs and the legspan of the largest species does not exceed six inches (15 cm). The most exaggerated rumors claim twice that, because they assume the six inches is body length and then double it to account for legs. I have seen some monsters from African and Middle Eastern desert areas and only the most spindly have had a 5″ (13 cm) legspan. The photos everyone has seen where a specimen or two are held by soldiers are doubtfully even that large. Except for a small group of highly adapted burrowers they have long legs and are built for speed. I have hand-collected and seen *Eremobates* in Arizona chasing down prey in nature and the reason for the name windscorpion is obvious—they really do move like the wind. They look like leaves blown rapidly across the ground on a particularly windy day. On close inspection the front pair of legs appears visibly thinner with reduced or missing tarsal claws, because they are not used for walking but serve a primarily sensory function. The sensory front leg pair is incredibly reduced compared to the ambulatory legs in some burrowing species. Despite the obvious function, the front legs are not referred to as

Ova from *Eremobates* sp., southeastern Arizona.

Eremobates sp. continued. Early 1st instars are pale white and possess long setae that keep them from being in direct contact with the substrate.

12 days later, extremities of the developing 1st instars grow pink before the first molt.

Ammotrechella stimpsoni from Florida. Note the large median eyes, massive chelicerae, and racquet organs. © Sam Floyd

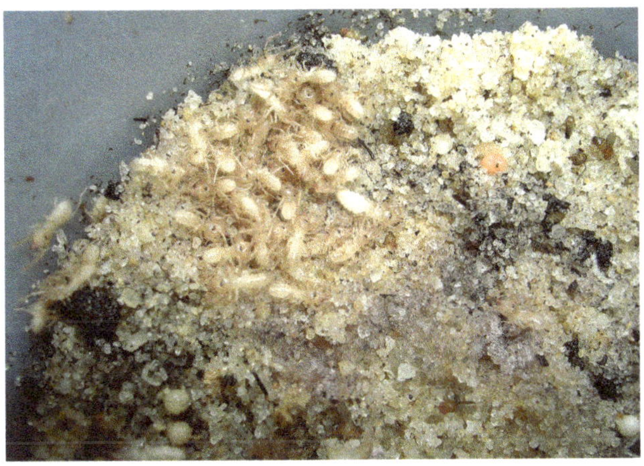

Eremobates 2nd instars are ambulatory and can feed but cluster together for a few days following the molt.

antenniform legs because they often contact the ground. On the bases of the rear pair of legs are unique, mallet-shaped organs called "racquet organs" which perform a similar function to pectines in scorpions. Males have accessory organs on the chelicerae or practice direct fertilization. Females dig a burrow and lay 40-200 eggs in a pile, but these rarely hatch in captivity and rarely continue past a few instars if they do. There is no visible postembryo stage (for *Eremobates* sp. at least) as eggs hatch into 1st instars that are functionally similar to the 1st instars of Amblypygi or Scorpiones. They have developed setae and appendages, are capable of movement, but are sedentary and do not feed until they molt to 2nd instar (McMonigle 2009a). Windscorpions have been seasonally available but they live only weeks to months, and captive breeding reports are exceptional at best. There are 1,113 species (Prendini 2011b) found primarily in desert habitats. Only a single species from India is known to possess venom glands, but they are not functional (Punzo 1998).

The Ricinulei are a group of about sixty small species also known as hooded tickspiders. They are brown and around the size of a tick, the body segments resemble a spider, while the "hood" is a plate that covers the chelicerae. They can detect light but are not known to have eyes. The second pair of legs are enlarged and used as feelers. The prosoma and segmented opisthosoma are joined by a pedicel. Males have special adaptations on the third pair of legs used to transfer sperm to the females, while the female lays one or two eggs and carries it around using her mouthparts

Hooded tickspider male.

(Beccaloni 2009). These are never cultured, primarily due to lack of availability rather than lack of size. They are mostly tropical but one species is found in Texas (Levi and Levi 1968).

Mites and ticks comprise the fascinating Order Acari with phases that swim, have suction cups, look like polyps, or have the wrong number of legs. They have just one body segment. Many mites are microscopic or difficult to see. The largest Acari are around the size of a pencil eraser with the exception of some ticks that can be much larger when fully engorged on the host's blood. The legs are usually short (shorter than the body for engorged ticks) but there are species with one pair of elongate sensory legs. Fertilization and development are varied but most females lay a large number of eggs that are abandoned. By sheer numbers unintentionally reared in captivity, they are the most popular captive arachnid, but they are pests, not pets. Some can be harmful parasites that live with or on humans and their pets and bedding. There are around 45,000 known species, thought to be a fraction of the real number, while infestations of a single species in a cage or the number of mites in your mattress might also exceed 45,000. Counts of over 1,000,000 individuals per square yard (800,000 per sq. M) are typical for temperate forest litter (R. and K. Preston-Mafham 2005).

Parasitic mites on a palpitore, Opiliones.

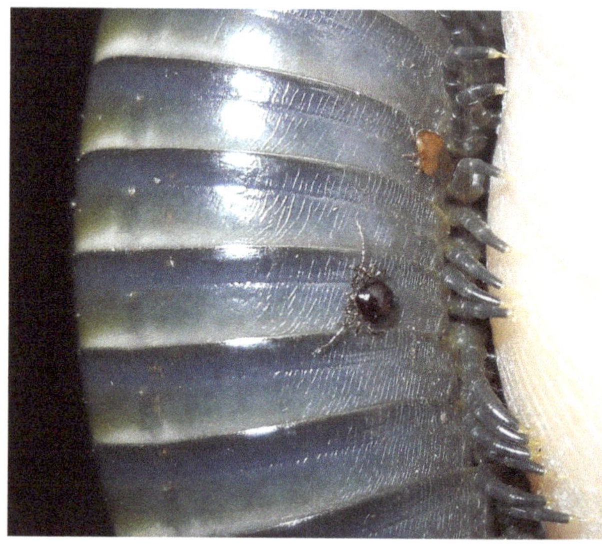

Some mites have a pair of elongate legs used as antennae like the specimen on the left. These two different commensal mite species spend their entire lives on a large millipede, *Acladocricus* sp., from Negros Island, Philippines.

2
WHIPSPIDER BIOLOGY

The Amblypygi are known for a flattened prosoma and uniquely specialized antenniform legs while the group as a whole consists of relatively large creatures. The very smallest (*Charinus* spp. and *Sarax* spp.) compare to the upper limits of mites, pseudoscorpions, harvestmen, etc. in body mass. The legspans are especially impressive as our smallest US amblypygid, *Phrynus marginemaculatus*, has a maximum front legspan of at least 4.5″ (114 mm). Almost every resource imaginable listing the largest arachnid debates which scorpion (including the tail) or spider legspan makes them the largest arachnid on earth (not even exaggerated and unsubstantiated reports claim greater than 12″ or 30 cm for any of these), yet not one mentions the whipspider. Even the highly dubious fossil spider *Megarachne* generated a firestorm of interest for its unbelievable twenty-inch (50 cm) legspan (Dunlop 1995) before it turned out to be a chunk of an aquatic sea scorpion. It may be because it is difficult to visualize or be impressed by a measurement such as 4.1 ounces ("This arachnid is so monstrously huge that an exceptionally fat, gravid female may weigh as much as one third the size of a small soft drink," is not seen) or a few inches body length, that attenuate, tail-like abdominal segments or ambulatory appendages are used for arachnid measurements. Certainly there are more massive arachnids and the largest spider (whether it be the South American *Theraphosa apophysis*, *Theraphosa blondi*, or the Asian *Heteropoda maxima*) is heavier, but legspan is the measurement used. There has never been contention over whether the whipspider's front legs are truly legs and yet there is never a mention of the documented 23.6″ (60 cm) legspan (Weygoldt 2000) of the gargantuan Amazonian whipspiders, *Heterophrynus batesii* and *H. longicornis*. According to

Whipspiders are not always flat.

Some species are more 3-dimensional than others, *Paraphrynus mexicanus*.

given measurement, these *Heterophrynus* are by a long shot the largest arachnids presently living on the planet Earth.

There are presently five whipspider families containing seventeen genera and 161 species.

Family Paracharontidae—*Paracharon*
Family Charinidae—*Catageus, Charinus,* and *Sarax*.
Family Charontidae—*Charon* and *Stygophrynus*
Family Phrynidae—*Acanthophrynus, Heterophrynus, Paraphrynus* and *Phrynus*.
Family Phrynichidae—*Damon, Euphrynichus, Musicodamon, Phrynichodamon, Phrynichus, Trichodamon,* and *Xerodamon*.

The Paracharontidae has a single, tiny member, *Paracharon caecus* Hansen 1921, found in termite nests in western Africa (Weygoldt 2000). The Charinidae and Charontidae have pulvilli at the end of the tarsi, which allows them to climb smooth surfaces, but so far none have been kept in the states. The Charinidae are tiny species found primarily in the tropics and few are half the size of our smallest native *Phrynus*. The Charontidae are generally small but one or more large *Charon* species from the Indo-Pacific region rival some *Damon* in measurements (Harvey and West 1998). The final two families lack pulvilli and contain most of the large species, including all those whose husbandry is discussed in this book. The Phrynidae are New World except for a single Indonesian species (Harvey 2002). Phrynichidae are primarily Old World with exception of the Brazilian genus *Trichodamon* with two species. Other than location, an easy way to tell these two families

Some species that live in caves have reduced eyes. The median eyes are very small on this *Charinus australianus cavernicolus*. © Michael Seiter

apart is by looking at the size of the small sternal plates (prosomal sternites) on the underside of the body between the legs. The small plates are reduced to specks in the Phrynidae. Also, the Phrynichidae palp has shorter tarsal segments and a large opposing spine on the tibia forming somewhat more of a hand.

The majority of amblypygids look very similar and are difficult to tell apart, especially as immatures. The arrangement of the tibial spines on the pedipalps is a primary characteristic used to identify genera, though most pedipalp differences are difficult to see on immature specimens due to reduced development and small size. The basal and distal articles of the chelicerae have rows of denticles that can be used in identification down to the generic level. The arrangement of trichobothria on the legs is also an important trait. In adults the shape of the female's dissected genitalia is another key indicator, while the male's is not. Color is not useful in identification because it often varies as much by age and proximity to the molt as it does by genus or species. The 2nd instar immatures of most or all species have red to orange pedipalps, while some have a green opisthosoma or yellow margins and stripes on the segments. In nearly all species the color fades to brown or gray after a few molts.

Although all amblypygids are dorsoventrally flattened and share overall anatomy, there are some notable differences in the overall body structure. The abdomen can be rounded or elongate at the posterior end. The pedipalps can be short

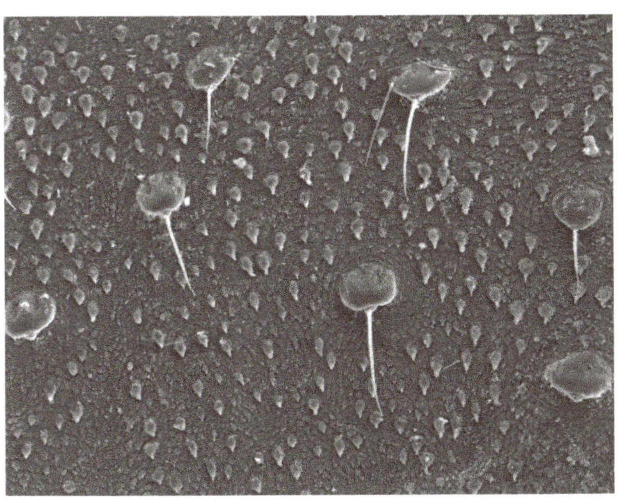

The intricate surface texturing and sensory setae on opisthosmal tergite, adult female *Phrynus marginemaculatus*, SEM.

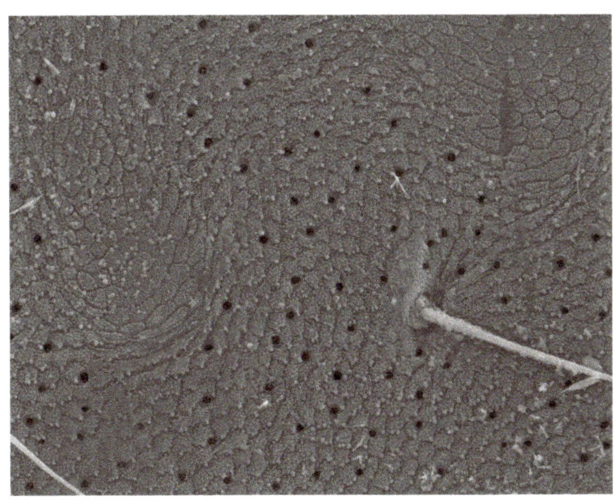

Sensilla on sternite, adult female *Phrynus marginemaculatus*, SEM.

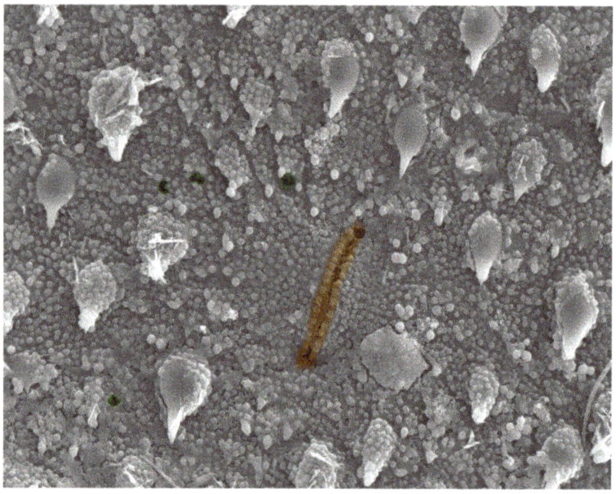

A probable slit sensilla (peach) on the opisthosomal tergite as well as small sensilla that resemble tiny stone plants.

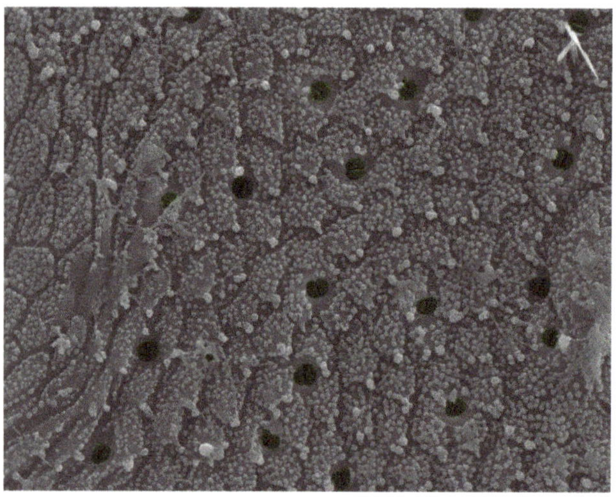

Sensilla on sternite, close-up, are similar in size and structure to those shown on the tergite, adult female *P. marginemaculatus*, SEM.

and thick, long and thin, or somewhere in-between. Some have elongate pedipalps covered with hundreds of large and small spines like *Damon medius*, while others like *Euphrynichus amanica* are nearly smooth, with just a few large spines forming the hand. Those with short palps can sport a massive catch basket of long spines like *Heterophrynus alces* or fat palps with a comb of teeth as in *Phrynus barbadensis*.

The second, third, and fourth leg pairs have long rows of trichobothria on the tibia just before the tarsus as seen here. The trichobothria and other large setae are relatively oversized on small immatures (this *D. diadema* is molting to 2nd instar at point of whip extraction by palps, it was pulled off mother's opisthosoma for photo).

The walking leg design is often flattened for sideways movement on a nearly two-dimensional plane and resembles the rapid, surface hugging movements of the sally-light-foot crab. However, a few genera, notably *Heterophrynus* and *Paraphrynus*, have extremely long legs that can hold the body away from the surface and make them look like an improbably large harvestman. A few amblypygid species that live deep in caves or ant nests have reduced or missing eyes, but since the eyes are not a notable feature, variations in ocular development do not result in a notable difference in appearance.

Every enthusiast should consider borrowing a stereomicroscope, if only once, to

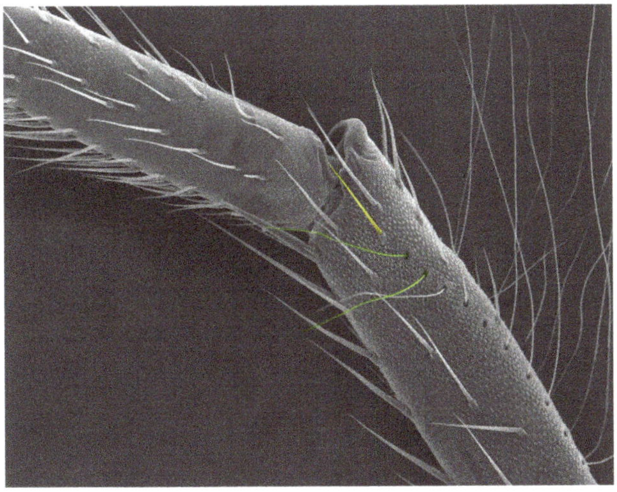

Tibia-tarsal joint spines and setae including the left row of trichobothria. SEM, *H. batesii* 3rd instar.

Close up of top two trichobothria.

get a close look at the whipspider. These are amazing creatures that show diversity few, if any, other plants or animals can match. Most single organisms are rapidly exhausted as the same structures and textures are repeated over and over again. Literally, hours can be spent focusing on the different structures and surface variations of a single amblypygid. Comparing different species or developmental stages can also be fascinating. My favorite structures may be the odd little pair of bulbous-based setae differently placed on the median and lateral eye mounds of *H. batesii* that look like tiny antennae on an old-fashioned flying saucer, though the pedipalps are always a work of art. More intense diversity can be seen at the submicroscopic level of the scanning electron microscope. The most interesting may be the great variety of sensory setae shapes and functions on the legs, but even the cheliceral combs display variation of form and shape.

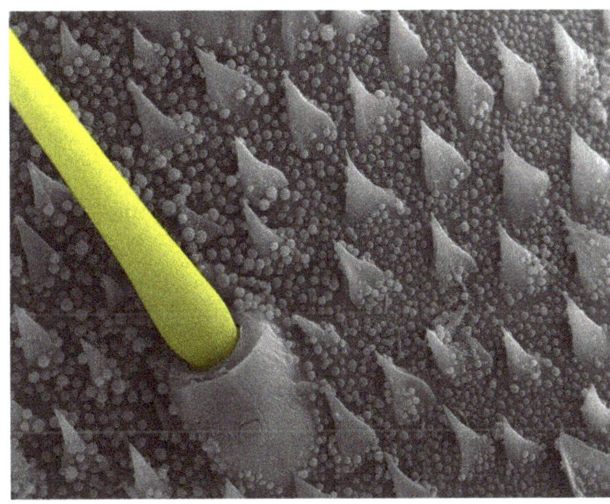

Close up of seta above trichobothria. There is even an unusual sensilla on the socket of this seta sensilla!

The Legs

The last three leg pairs are commonly called walking legs since they serve an ambulatory function. They can move the creature in any direction but are only built for rapid sideways movement and much slower forward or backward locomotion. Specimens can quite literally move to the other side of a piece of bark or the other end of the enclosure in a blink of the eye. Walking legs may also be flexed for spring-loaded jumping or leaping.

The modified front legs are unique and are unrelated to ambulatory speed because they are not used for weight bearing. Also

Walking leg claws (gold), SEM, *Heterophrynus batesii* 3rd instar.

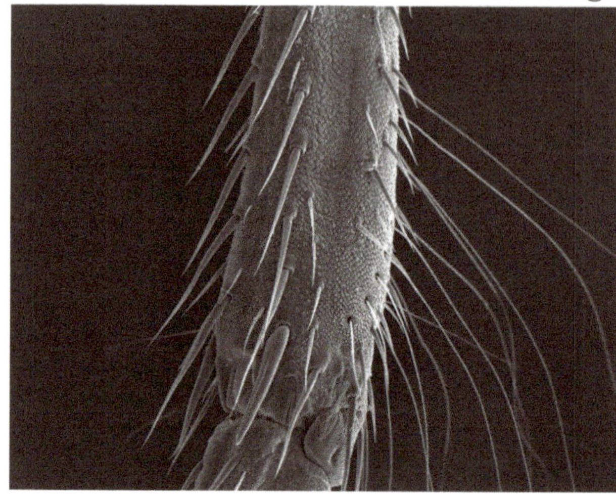

The setae and trichobothria of this adult female *P. marginemaculatus* have much in common with the tibial-tarsal joint of *H. batesii*.

Trichobothria on walking leg tibia, exuvium *H. batesii* 4th instar. The bilateral, triangular formation of pits is somewhat visible in the arrangement of circular discolorations of the exoskeleton. Even the longest filament shown here is only 2-3 microns in diameter but can be greater than 2,000 microns in length.

This is also the tibial-tarsal area of the same specimen at left, but of the antenniform leg. The triangle and bilateral rows of long trichobothria found on the same segment of the walking legs appear nonexistent due to extreme reduction in filament length and quantity.

known as antenniform legs (or "whips") they function to relay information about the surroundings. These appendages are far more versatile than the antennae of insects and crustaceans in form and function. The whips can feel around corners, investigate the opposite side of objects, initiate mating, be amputated in defense, or fold up out of the way during running and hiding.

The span of the antenniform legs is often not referred to as a legspan, but rather is labeled "whipspan" to differentiate it from the span of the walking legs. The maximum walking leg span (measured straight across) is probably the 11.5″ (29

Antenniform leg distal segments SEM, *H. batesii* 3rd instar (from shed exoskeleton).

Club sensilla used for olfaction are colored blue, vestigial claw red, while remaining setae are more variable than they appear here. On closer magnification the longer setae (light purple) have large sockets and are outfitted with even smaller fringes while some of the shorter setae are porous and likely chemosensory. Antenniform leg distal segments SEM, *H. batesii* 3rd instar.

cm) of some larger *Heterophrynus* species (measurement of *H. elaphus* specimen on public display (Greer pers. comm.)), but the documented maximum span of the incredible antenniform legs is known to reach at least 24″ (61 cm).

The legs provide incredible speed and sensory input but are also key identifying features for this order. The legs are similar in construction to most arachnids: coxa, trochanter, femur, patella, tibia, and tarsus. However, the patella on all eight legs is small compared to most members of other orders (and especially dissimilar from the elongate front tibia of the whipscorpion), while the tibia is normally broken into two or more articles, even for the walking legs. (For example, one specimen of *H. batesii*: 1st tibia 36 articles, 2nd tibia 2 articles, 3rd tibia 2 articles, 4th tibia 3 articles.) Extremely long trichobothria in two rows of a dozen surround a central triangle of three located on the last or distal article of the tibia, just before the tarsus, on all six walking legs. The linear pits converge towards the top so the structure resembles a stitched, upside-down triangle from the front. This inverted triangular patch is a unique feature of the Order Amblypygi. The same article on each whip has just a few, very short trichobothria. These and other short trichobothria on various parts of the legs can be difficult to identify even with a good stereomicroscope. When the filament is not long, the pit may be a different color from the exoskeleton and the setae are usually white, whereas other leg setae tend to be reddish or gold.

The Whips

Although the trademark antenniform front legs serve a sensory function like those of

Commonly kept species have no pulvilli and cannot climb smooth container walls.

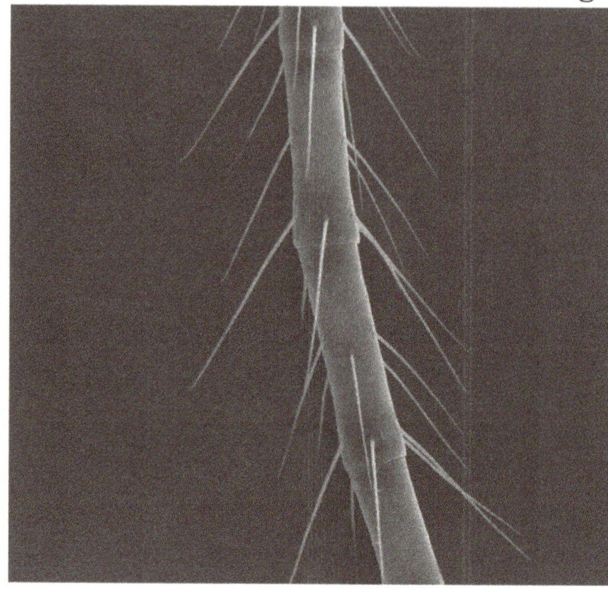

Antenniform leg tibial segments SEM, *Heterophrynus batesii* 3rd instar.

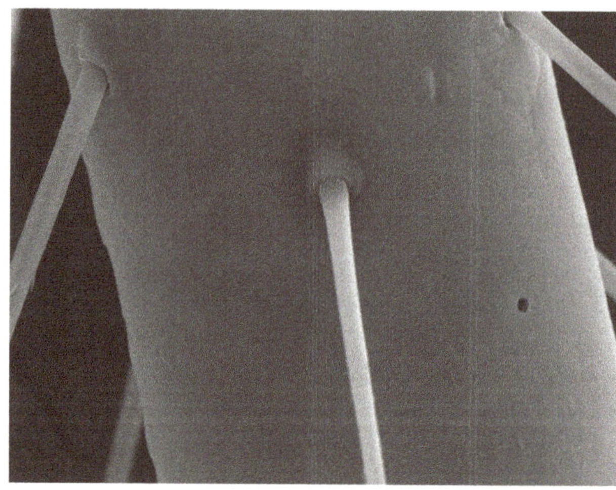

On closer magnification some small fringes can be seen on left setae and a tiny sensilla towards right of shaft. Antenniform leg tibial segments SEM, *H. batesii* 3rd instar.

a variety of other arachnids, they are many times longer and advanced segmentation begins after the patella. The tibia and tarsal segments combined are broken up into more than forty articles. A count of one 4th instar *H. batesii* whip yielded 77 tarsal and 36 tibial articles. Weygoldt (2000) reports that a single whip can have as many as 148 joints. Not only more complex, they are outfitted with as many as ten different types of sensory organs (Foelix and Hebets 2001) including lyriform, plate, and pit organs, slit sensilla, trichobothria, porous, rod, club, and leaf-like setae, which are used to detect texture, movement, shape, physical stress, humidity, vibration, odor, and air movement. The tiny sensory organs depicted in various SEMs included here may be an eleventh type: campaniform sensilla. One of the sensory organs is made up of microscopic vestiges of the tarsal claws as there are no actual claws. The lyriform organ on the trochanter and tibial trichobothria are miniscule compared to similar structures on the walking legs. Some of the types of sensory organs measure in the hundreds and each connect to the central nervous system by up to forty axons, so the interior is mostly nerves and axons. One actual count reported by Foelix and Troyer (1980) was 23,050 nerve axons at the joint between the tibia and tarsus. The front legs

The full length of the whips can be difficult to grasp on a live specimen because of three-dimensional movement and depth. Both of these legs are the same length and the tibia is shorter than the tarsus but it does not look that way in this photograph. Male *Heterophrynus batesii* from Peru.

are slow and methodical as they bend around obstacles, but they quickly fold up like an accordion for hiding and running. These "folds" and slightly enlarged segments can be the only way to differentiate the original tibia and tarsus. Each article is often similar in size and shape to the tiny adjacent patella, so the difference between the patella and first tibial article can be discreet, except in the case of autotomy because the patella is not lost unless the femur is.

The most significant antenniform leg functions are hunting and mating. Prey items can be located, measured, smelled, and possibly even tasted before the raptorial palps are called into use. Communication during mating involves various movement of the whips and it seems highly unlikely either gender could perform courtship displays if both whips were missing.

Whips may even be used to communicate through sound. Recent experiments have shown that whip vibrations appear to be a case of near-field sound communication perceived by sensitive trichobothria on the walking legs, particularly the two nearly microscopic trichobothria on the patella of *P. marginemaculatus* (Santer and Hebets 2007). This research related to male on male combat but probably relates to initial introductions and courtship as well. This may be seen as rapid whip vibrations when the whips do not touch the other animal. Therefore, the introductory whip flicking may relay sound more than touch.

The antenniform legs contain giant neurons which are found in other animals as a means to speed up escape response. Since non-ambulatory legs cannot be used to facilitate escape, giant neuron function is uncertain. One theory is giant neuron function may simply be required for rapid escape response of the antenniform legs themselves, rather than the whole animal. Without the specialized neurons it might not be possible to fold up the legs, out of the way, in a fraction of a second when the animal is ready to run off. However, the whip extremities have little to no muscles and are reined in by tendons. (This may be hard to believe when the intricate, 360° manipulations of the tarsal ends are observed.) The giant neurons receive input from tactile setae on the surface and are likely used in prey location or intraspecific communication. They are connected by giant nerve fibers to the central nervous system, not muscles. Giant neurons are thought to serve a sensory function though research has not yet proven exactly what purpose the giant neurons actually serve (Foelix, Wang and Beck 1975, Foelix and Hebets 2001, Foelix, Troyer, et al. 2002, Spence and Hebets 2006). The giant neurons are one possible explanation for why the antenniform legs do not always have to touch an object or other animal, but still seem to relay information. Such an ability to detect slight deformations of nearly undetectable electrical fields is a stretch. The large number of chemosensory setae that detect odors may be a more reasonable explanation, though proximal odor on such a level is uncertain and infinitesimal resistance to air movement of stationary objects is also an unlikely possibility. In the end, giant neuron function may simply be a way to provide rapid sensory information to the central nervous system that can be used to move one whip, both whips, or the entire animal in response to a particular situation.

Paraphrynus laevifrons late instar immature running wild in Costa Rica and missing a whip. It was seen under a gigantic rotten log that bridged a gap over a stream bed. This picture was taken below the log against a mud wall. Two other animals that appeared to be the same species and size were also below this side of the log. © Matt Kaplinsky

Memory relies heavily on input provided by the antenniform legs, though animals missing whips may still be able to find shelter, adapt to hunt prey differently, or recognize conspecifics with limited success. According to research in neuroanatomy, amblypygids have the largest mushroom body (area of the brain used for learning and spatial memory) of all the arachnids (Strausfeld et. al. 1998). Whipspiders recognize and remember each other for a time (discussed in the mating chapter) though this may not be possible without the antenniform legs. The exact length of time they remember each other is hard to gauge and would make an interesting study. *Phrynus marginemaculatus* have been shown to run mazes (whose prize is shelter, not food) more rapidly when given tactile cues with different textures to aid in finding the correct path, or incorrect path, when later manipulated by the tester (Santer and Hebets 2009). Specimens without antenniform legs were not tested because previous experiments (Beck and Gorke 1974, Weygoldt 2000) revealed that animals after having the whips removed could not find their way back to shelter.

Whip grooming is a common behavior.

Other than for locating prey, sizing up competition, courtship, and finding shelter, the antenniform legs are used to escape predation through autotomy (like the familiar tail of certain lizards or legs of certain walkingsticks). It may seem counterproductive to throw one's most important sensory appendage at a predator (like plucking an eyeball out and throwing it at the threat) but whips are extremities and abandoned only in the case of life or death in the wild. Other invertebrates voluntarily give up large amounts of blood or significant ambulatory appendages in hopes of escape. In captivity, adults would require the whip to be pulled on while 1st instar whipspiders may seem to autotomize with reckless abandon. Escape from a predator is not the only function. Captive specimens have autotomized after minor damage to the whip, when the very tip is trapped under a lid, when whips are trapped during or deformed from a bad molt, and due to stress. Differences in autotomy relative to age and the point of damage are discussed in the health and regeneration chapter.

Amblypygids are active when there is very little light and use the incredibly sensitive antenniform front legs to "see" the surroundings, but they are not helpless without them. One missing whip causes no change in behavior. The front tibia of the walking legs especially, and the palps, have very sensitive, specialized hairs located in individual sensory pits called trichobothria. Along with a variety of other large setae, they are able to detect vibrations, chemical cues, and slight air movements. These help alert the animal to suitability of the prey and potential threats but are secondary to the whips. Sensors on these appendages easily make up for a single missing whip in the wild. If both whips are lost they will adapt to using pedipalps and sensitive trichobothria on the second pair of legs to hunt until they can regenerate their whips. Minor vibrations resulting from a gentle tap to the habitat or the carefully slow opening of a lid can cause rapid appendage retraction or flight responses even if both whips are missing. Beck and Gorke (1974) observed that animals missing both whips were unable to locate their shelters or catch prey. Yet, in captivity, specimens missing both antenniform legs are able to capture prey without difficulty and do not require shelter. In nature where predators lurk, scarce prey must be hunted, and intraspecific communication can reduce cannibalism, two missing whips is probably a death sentence.

Damon diadema eyes.

Heterophrynus alces eyes.

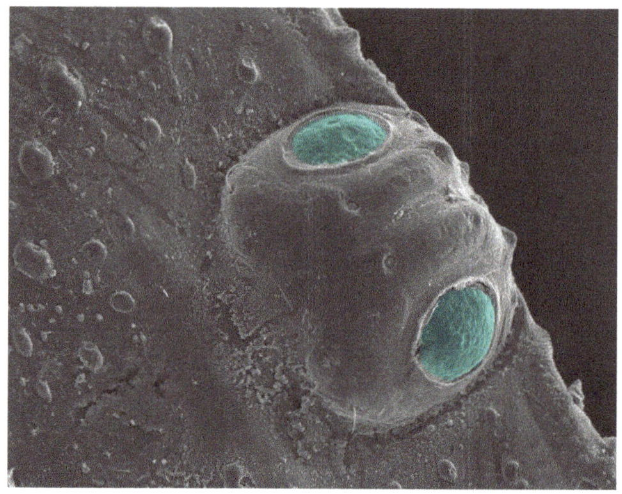

Median eye mound of an adult female *Phrynus marginemaculatus*, SEM. The surrounding, raised plate-like area of the prosoma holding the eyes is sometimes referred to as an ocularium.

The long whips may be very important but they can also give away an animal's presence. Small immatures of known species all have the distal tibial segments and the first tarsal segment white in color. (Species with incredibly long whips like *H. batesii* have an additional white area on each side of the femoral-tibial joint.) This breaks up the outline of the whips so they are not as easily seen on a light colored or patchy background. A number of insects have white bands on the antennae for the same reason. Large immatures and adults grow out of this feature since large white patches on the thick, brown whips would make them more conspicuous.

The Eyes

Amblypygids usually have eight eyes. The two central eyes are often the largest and are located on a mound on the front of the prosoma just behind and between the chelicerae; these central eyes are found in a number of arachnid orders and are also known as median eyes. Two lateral clusters, each with three eyes, are located a little further back on the prosoma to the outside of each chelicera. In each lateral cluster the inward facing eye faces up and is a different shape, while the other two face sideways, roughly similar to the fore and aft angles of the adjacent median eye. The inward lateral eyes are white in color on living specimens, while the remaining six are black. All lenses appear white on exuviae. It is difficult to show the three-dimensional (or lack of) aspects of the eye

All eyes on an exuvium appear white because only empty lenses remain. *Heterophrynus batesii* 4th instar.

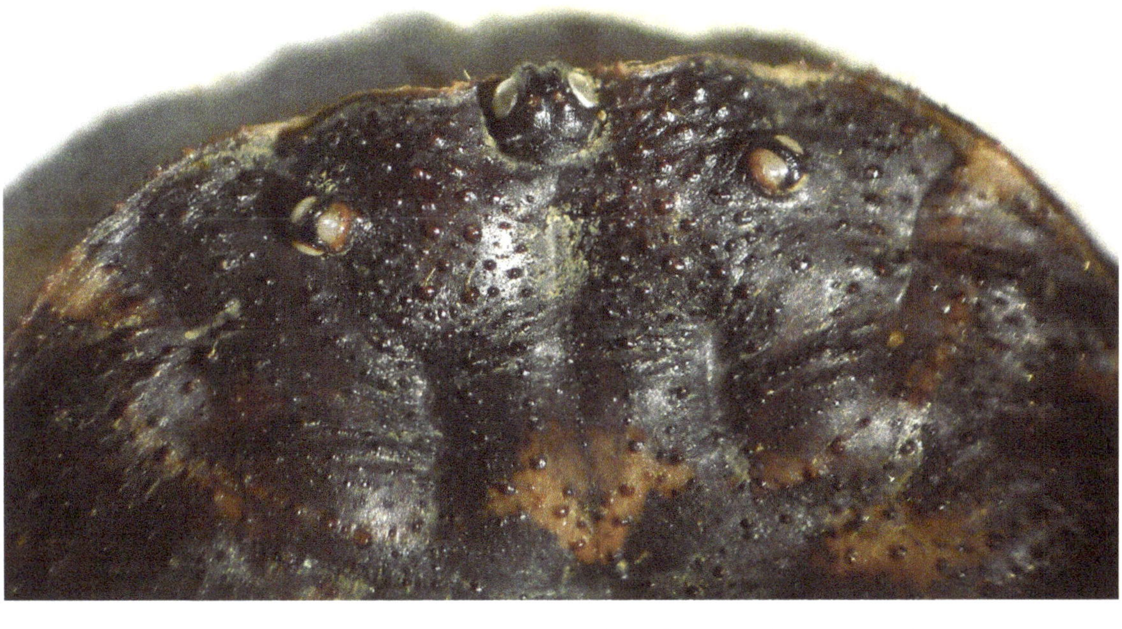

Damon diadema eye structure, exuvium. Perhaps the eyes on the back of the head are what make the female such a good mother?

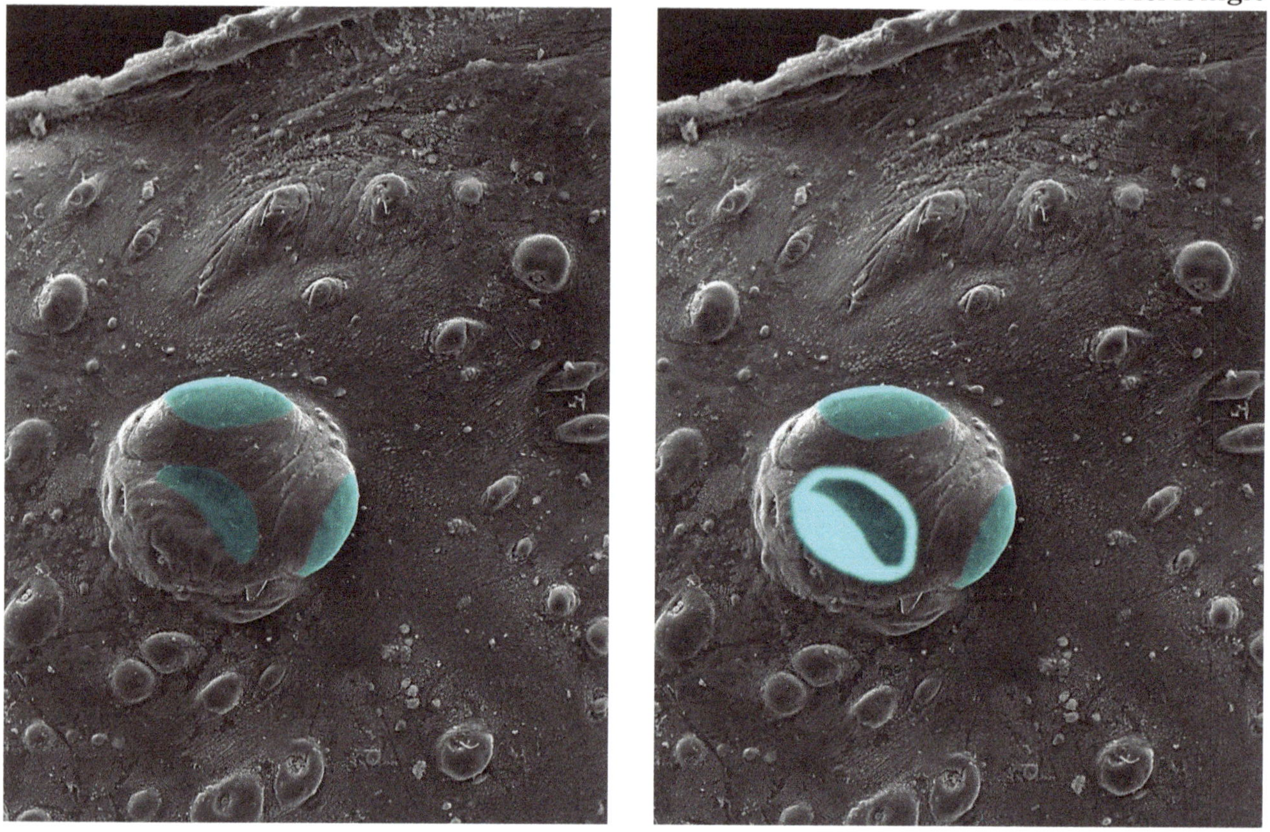

Right lateral eye mound. The large, white-colored, inner eye shows up as a small, dark, barely distinct lens (left) because SEMs pick up texture, not color. *P. marginemaculatus.*

mounds in photographs. In some species the median or lateral mounds barely rise from the surface, while in *Heterophrynus* all three mounds look like tiny turrets from the front or side. Conversely, the median eye mound can be located in a deep, protective pit as in *Phrynus barbadensis*. Usually the eyes on the central mound are almost touching, but in some species like *Phrynichus exopthalmus,* the pair is relatively far apart. Cave species like those pictured in the text usually have all eight eyes, sometimes with the median pair reduced, but there are a few species, such as *Charinus quinteroi*, found in ant nests, that have just six eyes (Weygoldt 2002)—the two median eyes are missing. Beyond those there are a few very rare, small species missing all eight eyes, such as *Paraphrynus reddelli* Mullinex 1979.

Although there is no question of the ability to recognize light and dark, the ability to see images is uncertain. Whipspiders rely heavily on other senses and respond

Damon diadema chelicera.

differently to sudden changes in light than to vibrations, so visual responses are more difficult to test. A light tap to the glass, puff of air, or touch of the leg commonly results in an obvious body shudder or rapid movements. If a bright flashlight is shone on an animal, it may close the palps slowly and freeze in place, but only if it was in the process of moving. Reaction to light is often delayed and the animal may simply move slowly out of the circle of light. Some of the large species have relatively larger eyes and appear to have some ability to recognize movement of shapes. *Damon* vision may be secondary (or tertiary) in prey capture or courtship. Specimens commonly throw their food away if the enthusiast's face gets too close to the enclosure (without casting a shadow), so they certainly can recognize more than simple light and dark. *Heterophrynus batesii* also display obvious reactions to my movement when a shadow is not cast, but as with *Damon,* these reactions normally occur only during feeding and are not easy to initiate. *Heterophrynus* move reactively, but while *Damon* stays frozen in place, flicks the palp, and tosses the food, *Heterophrynus* moves away and hangs onto the

The chelicerae close down against the coxae of the palps. The palpal distitarsi would fold in between were this a live specimen, *H. batesii* 4th instar exuvium.

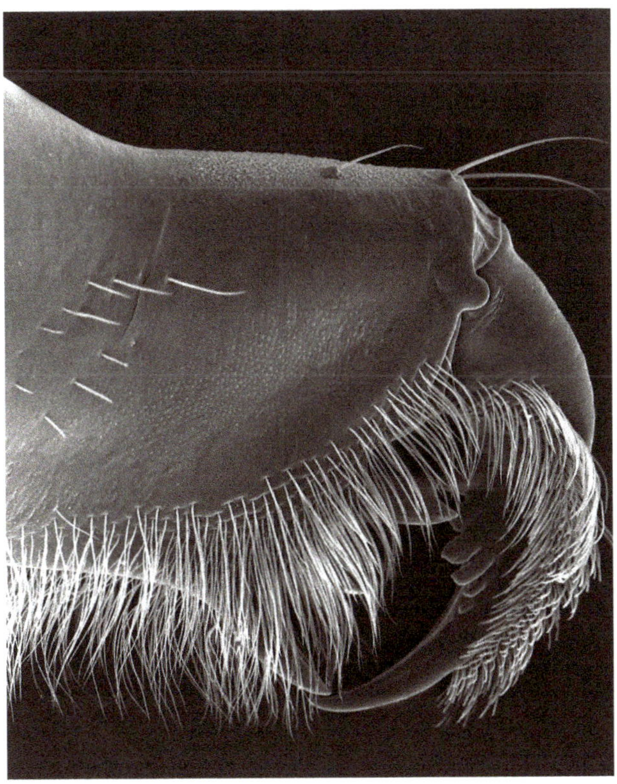

Inner surface of a chelicera, SEM, *Heterophrynus batesii* 3rd instar.

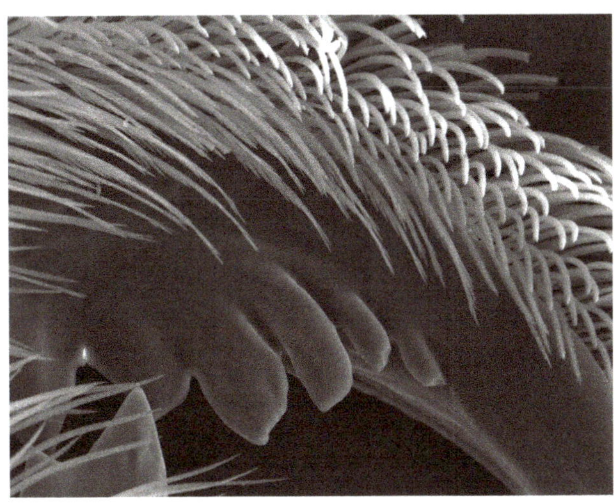

On closer magnification areas of feathery setae and curved brush-like setae are visible.

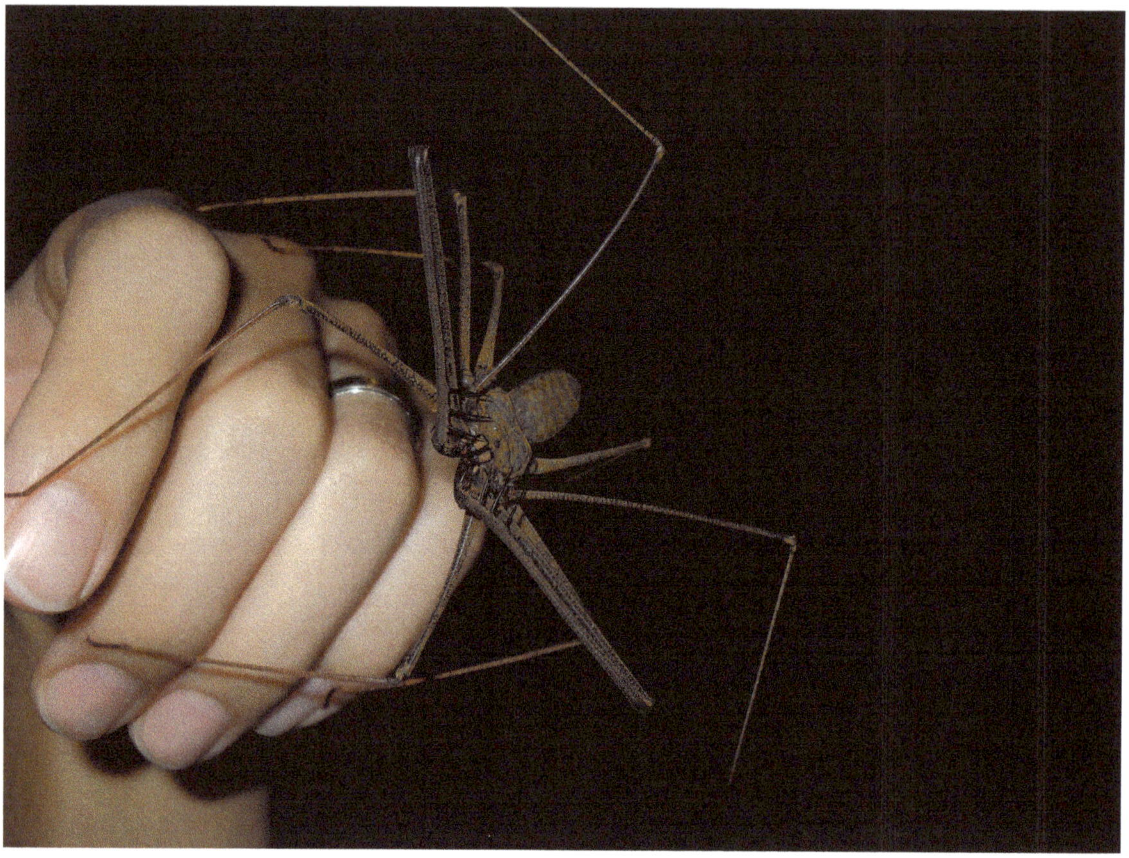

Whipspider palps. (With such a 3-D creature the depth and size do not translate correctly.)

prey item. Both genera display some courtship behaviors (palp movements) the other animal cannot feel, but might be seen with rudimentary shape recognition. Also, when prey is tossed in on the other side of the cage, they often direct a whip towards it in advance. Visual detection of movement is a plausible explanation.

Notice the long chelicerae and a parasitic mite at the base of the antenniform leg (left), *Phrynichus* cf. *phipsoni* from India. © Dan Wulf

The Chelicerae

The chelicerae are the smaller pair of mouthparts that chew up the food. In the Order Amblypygi they always face downwards and are composed of only two segments. The large basal segment usually has three big teeth, while the small segment has a row of much smaller teeth. The inner and outer teeth can vary greatly by genus and

Damon diadema incomplete palpal hand. Tarsal segments do not form a pincer but somewhat of a hand is formed by large spine on the tibia facing a large spine on the basitarsus.

are useful identification features on dead specimens or exuviae. The chelicerae to not contain silk glands nor venom glands, but they are used in fighting and are the primary tool used to cannibalize or inflict damage on other whipspiders.

The standout cheliceral feature is the thick mat of orange setae which ensures the animals do not ingest large particles and aids in grooming. Setae on the basal segment are long and smooth. The brush of "hairs" on the distal segment, or fang, includes downward-curved setae that are long and feathery or whose lower surface is outfitted with fine, brush-like fringes along each side. These aid in the important task of keeping sensory apparatus clean. Chelicerae are regularly used to groom the walking leg and palp tarsi, and a good portion of the whips (such extensive grooming is usually not associated with arachnids). They even pull the middle and back legs forward to "lick" the tarsi clean. Cheliceral setae must also be good conductors of sensory information as they are used by the male to check the structure of his spermatophore.

Euphrynichus bacillifer female with 1st instars. Note the nearly spineless palps, the hallmark of this genus. © Michael Seiter

The Palps

The most spectacular and menacing aspect of the amblypygid body is the unique pair of huge, spine-covered palps. "The pedipalps are the main distinguishing characteristic of the order" (Kirby 1986). The Amblypygi, specifically *H. batesii*, likely holds the record for world's greatest arachnid palpspan, though I have seen a few *Heterometrus* scorpions that might not be more than an inch off. These appendages are a requirement for fighting and feeding. Palps are the singular prey-catching mechanism and even short-palped amblypygids would look particularly unimpressive without them. In captivity, an animal missing both whips and a few legs can do perfectly

Pedipalp tibial spines, SEM, *H. batesii* 3rd instar.

Distitarsus and basitarsus, SEM, *H. batesii* 3rd instar.

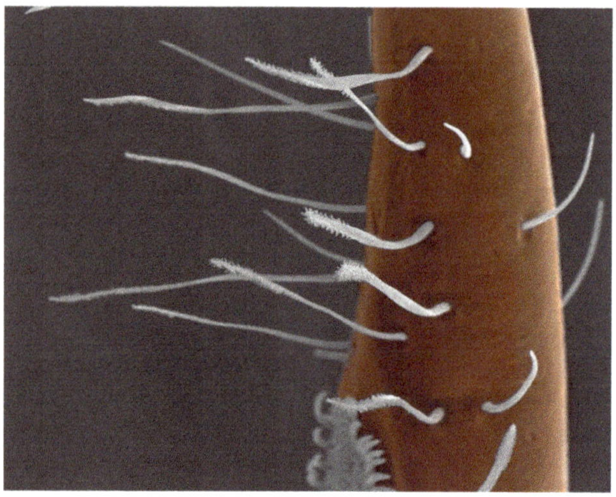

Distitarsus, specialized fuzzy setae, SEM, *Heterophrynus batesii* 3rd instar.

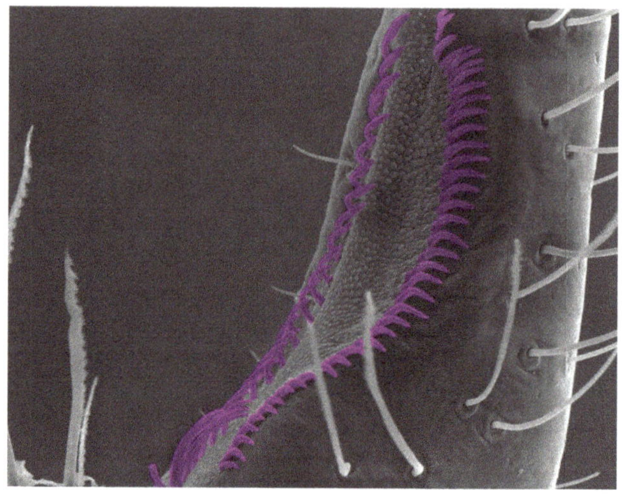

Distitarsus, rows of short comb setae used in grooming, SEM, *H. batesii* 3rd instar.

fine, while an animal missing half of a single palp is severely handicapped or doomed. Though the supremacy of this feature could be argued since they are never called "spiny palpspiders" it is difficult to argue against the commanding presence of the massive palps.

The spiny, raptorial palps are similar in segmentation to the legs but lack a patella and the tarsus has only two segments. They open up into improbably large arms that have a spiny, incomplete hand and two tarsal segments (the large, spined basitarsus and much smaller moveable claw known as the distitarsus). The tarsal segments can oppose the tibia somewhat, but not each other, and so do not form a clear pincer as in whipscorpions. The appearance is even less like the opposable pincer of scorpion and pseudoscorpion palps. Also, the whipspider palpal tarsi fold back against or between the chelicerae at rest, very unlike the tarsal pincers of these other arachnids that fold together or close at rest.

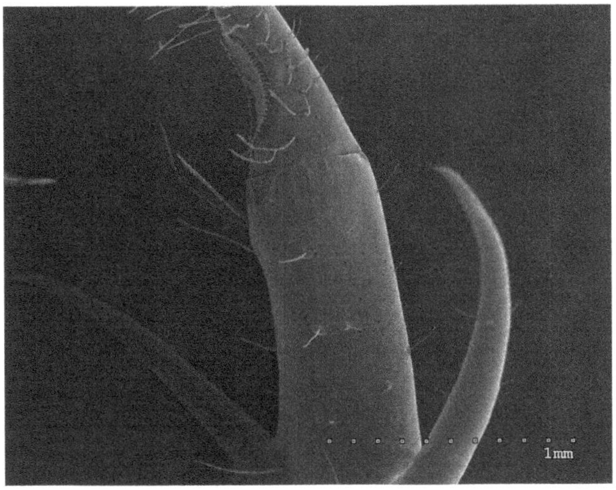

(A) Palpal tarsus with easily identified sensilla colored in. A number do not show up in this distant view.

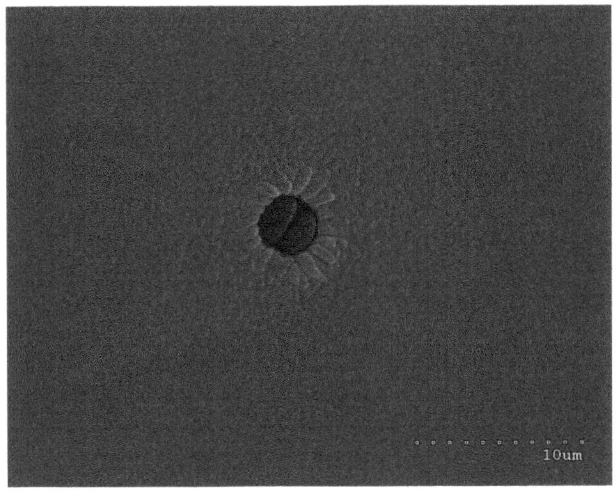

(C) Sensilla 2nd over and 1st down from seta in B. Submicroscopic sensilla like these (dark green) are arranged in such a way on the palp to suggest taste receptors, but similar looking sensilla are found on the sternites, tergites, antenniform legs, and walking legs in differing arrangements as shown. SEM, *H. batesii* 3rd instar.

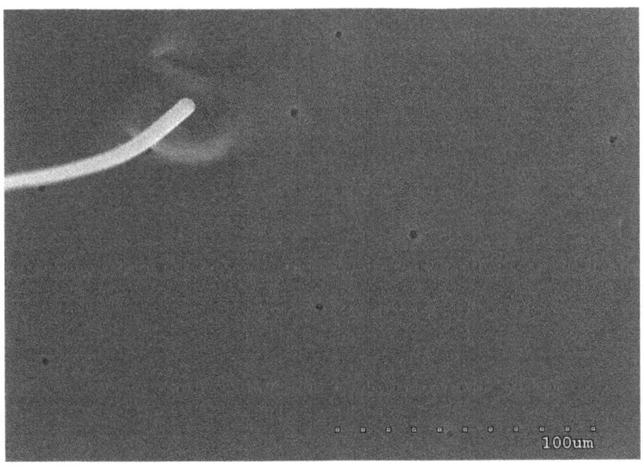

(B) Top left seta of basitarsus from A.

The very different palp structure is likely why whipspiders cannot burrow. (Pseudoscorpions, scorpions, and whipscorpions are excellent burrowers.) The stiletto-like distitarsi not only impale prey but also have complex, fuzzy sensory setae, and two comb-like rows of curved setae that play an important role in grooming.

Genera from *Damon* to *Phrynichus* and *Phrynus* open the palps along the same plane of the body, or at a slight angle if palps are overly long or in order to feed close to the surface. They capture prey with a deadly hug by pinning food between the two palps and pulling it close. *Acanthophrynus, Heterophrynus,* and *Paraphrynus* open the palps on a flat plane during feeding, but for hunting they rotate the arm-like palps perpendicular to the body, mirroring the position of the chelicerae. This allows them to pin prey against the ground or back against the chelicerae. However, even for these species the palps normally rotate back into position during the last movement of the capture swoop or during feeding.

Most species rarely, if ever, exhibit a threat display to humans, but some large species menacingly open their palps and stand their ground. The deimatic display can be the result of anatomical or behavioral differences, but the primary body part involved is the menacing, spike-covered palp. *Heterophrynus alces* has an intimidating threat display as it lifts its body,

1

2

3

4

Molting Sequence (opposite):

1. The prosoma and opisthosoma are extracted while weight is supported primarily by the extracting femora.

2. Once extracted, the tail at the end of the opisthosoma grabs onto the exuvium as evidenced here by deformation of the exuvium just before the genital operculum.

3. Leg extremities are extracted as nearly all weight is shifted to the tail segments.

4. As completion nears only the tips of the antenniform legs remain in the exuvium and are carefully extracted by the palps. If the tail were not holding the animal in place the attenuate whips would tear right off (since the animal cannot float in mid air).

(Above)
This shows *H. batesii* 5th instar coloration and dimensions approximately three hours post molt. Notice the extreme increase in the size of the prosoma, legs, and palps compared to the 4th instar at left. After a week the yellow prosomal margin becomes less visible at this instar.

rotates the pedipalps ninety degrees, and holds them back up over the prosoma, similar to the hunting posture (McMonigle 2011). It commonly exhibits this behavior in captivity and can follow it up with sharp, penetrating jabs. *Heterophrynus batesii* is also outfitted with large spines and even more impressive palps in late maturity, but it does not display the defensive behavior of its relative. *Damon diadema* are known to exhibit a deimatic display but this is almost never observed in captivity. However, Rayor (2007) reports females protecting offspring sometimes raise their body up and open the palps widely, while slowly advancing towards the threat. *Damon tibialis* is similar in structure to other large *Damon* but it has a pronounced behavioral threat display. The palps are opened wide as it bites at the air and waves the whips

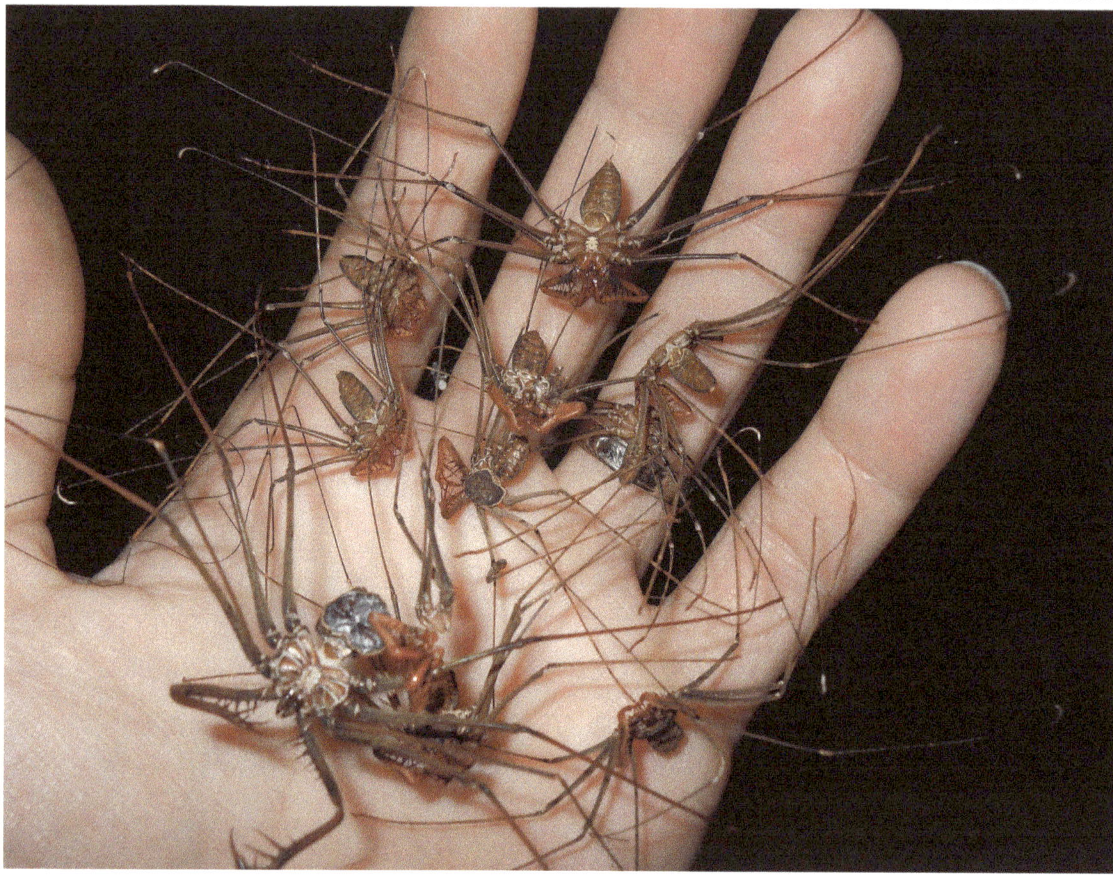

Exuviae.

wildly (Seiter pers. comm.). Small species occasionally appear to exhibit similar behaviors to cage mates or overly large prey but never towards the keeper. At least two whipspiders, the ultra-rare *Musicodamon atlanteus* and the massive *Acanthophrynus coronatus*, can stridulate during the palp display to try to scare off large predators and humans. According to Shear (1970), *A. coronatus* sounds like a small rattlesnake. The noise is made by rubbing together rows of paddle-shaped spines located on the inner surface of the chelicerae.

The palps are also used for drinking water. They do not usually lower the chelicerae down to a water droplet and drink like most arachnids and insects. The palp tarsi (distitarsi are each outfitted with long,

Heterophrynus batesii molt to 4th instar. The specimen is still hanging by the tail but the tarsi are just starting to test out the surface as palp extraction is being completed. (Disturbance can cause a freshly molted animal to flee prematurely.)

fuzzy setae and short combs used to aid chelicerae in grooming that can also hold small water droplets) are placed in water droplets and then "licked off" with the chelicerae. Other invertebrates may drink while grooming but do not use it as the primary drinking behavior.

Replacement of the Exoskeleton

Like all arthropods, whipspiders must periodically shed and replace the hardened plates that make up the exoskeleton in order to grow. Besides proteins, the main component of the exoskeleton is chitin, a complex sugar structure similar to cellulose, which is difficult to break down and so must be replaced for growth. The shed exoskeleton (called exuvium, pl. exuviae) includes the palps, legs, prosoma, tergites, sternites, and intersegmental membranes, as well as the lining of the book lungs and spermathecae (when present). This process, known as molting, provides a new suit of armor but is also an extremely vulnerable transition.

Signs of the impending molt are seen long before the molt takes place. Up to a few weeks prior to the molt, feeding stops and the opisthosoma rounds out (it can actually grow fatter without feeding). During the two days leading up to the molt, the amblypygid is commonly seen hanging upside down with all the appendages fully outstretched and frozen in place. It almost seems to be stretching in preparation.

When ready, the leading edge of the prosoma splits and the animal begins to extract itself using blood pressure over a fifteen- to thirty-minute period. Low humidity prior to molting can reduce body

Heterophrynus batesii 4th instar coloration after four or five hours.
The tiny opisthosoma will dwarf the prosoma after a few weeks of feeding.

Damon diadema a number of hours after the molt.

Heterophrynus batesii molt to sexual maturity thirteen months post dispersal.

fluids to a level below that which is required for successful molting, so increasing humidity or moisture after a molt has already begun is useless. The prosoma and opisthosoma are removed entirely first and the small tail appendage at the end of the opisthosoma grabs onto the inside of the exuvium just before the genital operculum. Gravity and physical extraction take over from blood pressure at this point. Weight is shifted to the tiny but important tail appendage and then the palps and walking legs are slowly pulled from the old shell. The extremely long whips are still only partly extracted when the rest of the appendages are free. The creature now hangs entirely by its tail. The teneral palps, now freed, are used to carefully extract the main body of the whips. Once all appendages have been extracted, the animal hangs from the tiny tail-like segments at the end of the opisthosoma. (The segments hold tightly to the opisthosomal exuvium which itself is

Freshly molted *Damon diadema* adult has a green opisthosoma.

This *Charon grayi* found under a stone illustrates animals search out a hiding spot after molting in the wild, Nalungan Island, Philippines. © Christian Schwarz

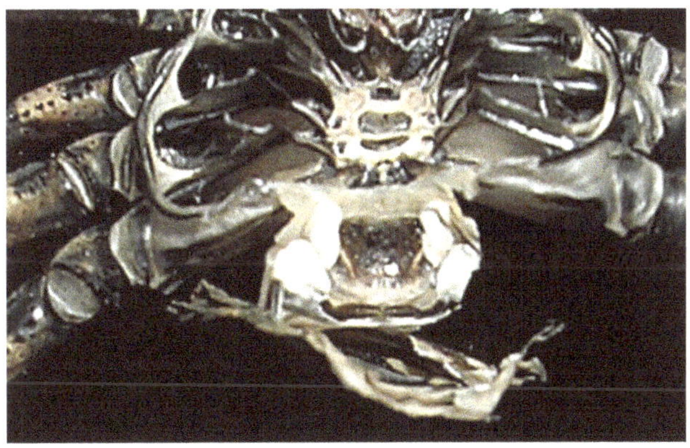

Exuvium showing book lung linings.

still held in place by the cast off tarsi.) The new exoskeleton is allowed to harden for up to half an hour and then the animal tests the surface with the feet—later the small "tail" finally releases its grip on the exuvium. The tail segments will release their grip prematurely if disturbed, which can lead to slightly bowed legs. The new exoskeleton is incredibly soft and pliable during the molt so any attempt by the keeper to aid the molting process will result in ripped off or stretched-out legs and then death.

Hardening of the exoskeleton takes much longer than the molt itself. The freshly molted animal is white though the opisthosoma can be green or white. Within a few hours to half a day, the exoskeleton darkens to blue or green, and then more slowly transitions to brown or black over the next few days. Powdery glandular secretions notable on some species take even longer and final coloration may require up to a few weeks.

The exuvium is never eaten and is usually knocked off by the animal's movements shortly after it finishes, or by the next day. Parts of old exoskeletons can be seen among the substrate in some of the included photographs since, while scavengers eat the intersegmental membranes, the plates and segments can sit around for decades.

LIFESPAN

The total lifespan for most species is uncertain but divided into two main parts: development and post-maturity. The time taken to develop does not add or take away from time as an adult unless it affects overall health. The molt to 3rd instar takes only thirty to sixty days with consistent feeding. Without adequate feeding, molts can be severely delayed. Some miniature species may mature as early as 5th instar but commonly kept species become sexually mature (or at least sexually capable) at the equivalent of 7th or 8th instar. Once sexually mature, the count starts over as adult stages. Later molts take longer and longer, until the adult molts that occur every year or two. The time taken to reach maturity is normally from twelve to thirty months though much longer is possible.

A large greenish *Paraphrynus raptator* adult. © Sam Floyd

In captivity the adults usually live two to four years after becoming sexually mature. Death of adult females during or shortly after gestation can be a common problem in captivity. (Some causes are discussed in the reproduction chapter.) Level

Mature male and female *H. longicornis* from Pará, Brazil. © César Augusto Chaves Favacho

of survival for reproductive females in the wild is unknown for comparison. Males may seem to live longer in captivity since they are not harmed by spermatophore production. So far no specimen I have actual data on has lived more than five years after reaching maturity (a female *D. diadema* that was not mated in its final four years), while three years is the most common post-maturity span. The total lifespan from 2nd instar to old age is normally between four and eight years. It is possible some individuals might live much longer, but those would be an exception rather than the rule. It would probably require minimal effort to keep a *Damon* spp. or *Heterophrynus* spp. for ten years in total by artificially extending the developmental period through metered feeding and avoidance of mating, if longevity were the singular goal.

3
THE SOCIAL WHIPSPIDERS

WHILE I MAY have read about them before in passing, I first remember hearing about these amazing creatures in 1993 from Barney Tomberlin of Hatari Invertebrates. He explained that these strange little arachnids (*Phrynus marginemaculatus* from Florida) had strange, incredibly long forelegs used like antennae and were predatory, but were normally kept communally because they did not eat each other. I was intrigued by the description and acquired some within days. I was permanently hooked when I first saw them and observed their alien appearance, movements, and interactions.

I kept captive-hatched immatures together in 1994 and afterwards, but their ability to cohabitate was well-known by arachnid keepers at the time. This was not the forced cohabitation of adult giant water bugs, various predatory beetles, and a few mantids that can be kept together at maturity only because they do not possess the voracity or strength to overcome a well-protected cage mate. Early instar whipspiders commonly consume prey larger than themselves and have the size and strength to easily overcome and cannibalize a sibling, but they do not, even when starving. (Compare this to the highly refined eusocial ants that massacre their sisters, especially larvae, at the first sign of hunger.) However, intermittent cannibalism reported for large immature specimens of *Damon diadema* (McMonigle 2005, Harrell 2006) contrasted with rumors suggesting greater sociality.

Proof of sociality in literature lagged a little behind its common acceptance in husbandry circles. *Phrynus marginemaculatus* were normally kept in groups at least as far back as the early 1990s. Intraspecific tolerance of sibling groups and pair behavior has been detailed in some husbandry literature relating to *D. diadema* (Ditzel 2002, Dixey 2002, McMonigle 2005) and *P. marginemaculatus* (McMonigle 2008 and 2009b). Common practice was to keep harem groups of adults—a single male with multiple females—with *D. diadema* (Dixey 2002), or single pairs (McMonigle 2005), while *P. marginemaculatus* were often kept in unsexed groups of a half dozen or more. Resulting immatures were reared in communal enclosures, at minimum for the first few instars. Weygoldt (2000) mentions pairing and communal behavior of some species but considered them primarily solitary animals. Rayor and Taylor (2006) demonstrated sociality through some experiments, but primarily by rearing up groups of young with their mother

Damon diadema mother and large immatures. © Ron Wagler

Some species, at times, will sit with the palps extended all day long. These individuals happen to be facing each other but specimens may face outwards or in the same direction, *Paraphrynus mexicanus.*

Phrynichus exopthalmus seem to be social here, but there is only one large cork bark surface in the habitat.

and documenting behaviors. The immatures were often found orienting themselves to the mother. Active aggregation and frequent tactile interaction were also quantified through all life stages for *D. diadema*, but only the early instars of *P. marginemaculatus* because, at nine months, their experimental groups all died from mites. The *D. diadema* were safely kept with their mother until they were large enough to be easily sexed through dimorphism of the palps. Arachnid social behavior, other than parental care of the first few instars, is extremely rare. Less than 0.1% of known arachnid species live in interactive social groups for extended periods (Rayor 2007). Before her experiments, Rayor had been told the mother would eat the young, but as of 2013 she has not experienced a single case of cannibalism by the mother in multiple groups kept together for one year following hatching (2013 pers. comm.). There has been little to no research on whipspider social tendencies in nature, though reports of cohabitating mature pairs are common. While Hebets (2002) found as many as 16 *Phrynus parvulus* on a single tree in Costa Rica, numbers were relative to the size of the tree and the animals were widely spaced around the tree. Weygoldt (1977) reported a single tree may host a pair as well as a number of immature *Heterophrynus longicornis*.

Walsh and Rayor (2008) documented that *D. diadema* have the ability to identify and behave differently towards kin (or at least familiar specimens, but the difference between the two was not investigated). At approximately the same time, *Culturing Gentle Monsters* reported related experiences

Figures on these two pages used by permission of Linda Rayor. From: Rayor, L. S., and L. Taylor. 2006. Social behavior in amblypygids, and a reassessment of arachnid social patterns. *Journal of Arachnology* 34: 399-421.

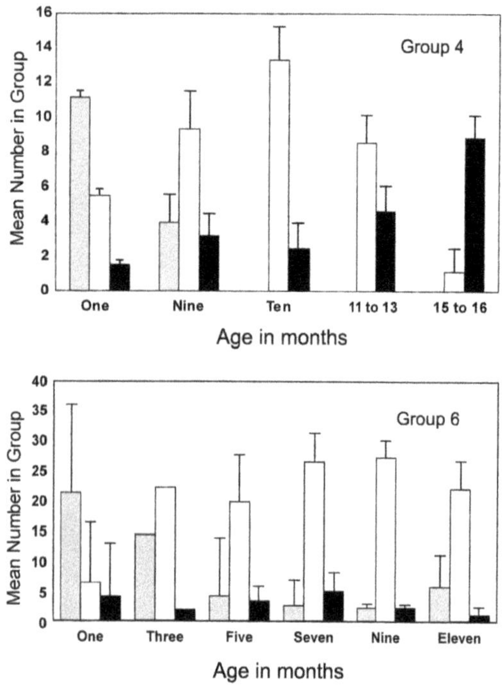

Mean number (+ SD) of *Damon diadema* young found in association with their mother (grey bars), siblings only (white), or found solitarily (black) at different ages. Group 4 was composed of 18 siblings and their mother until they were 10 months old, when two subadults died of natural causes. The adult female was removed when the offspring were 11 months old, due to her increasingly aggressive interactions with her offspring. Between age 13 and 15 months, the amblypygids reached sexual maturity. By the time the siblings were 15 months old, 7 siblings died after losing appendages during aggressive interactions or starved while evading their more aggressive siblings, resulting in n = 10 adults. Mean numbers per group were based on 8-15 observations per month. Group 6 was composed of 38 siblings and their mother. All siblings were present until age 11 months when eight subadults died. The mother remained with her offspring throughout these observations. Mean number of animals in each category per month was based on 3 or more observations (range n = 3-10 observations/month), except for when the animals were 3 months old (observation n = 1).

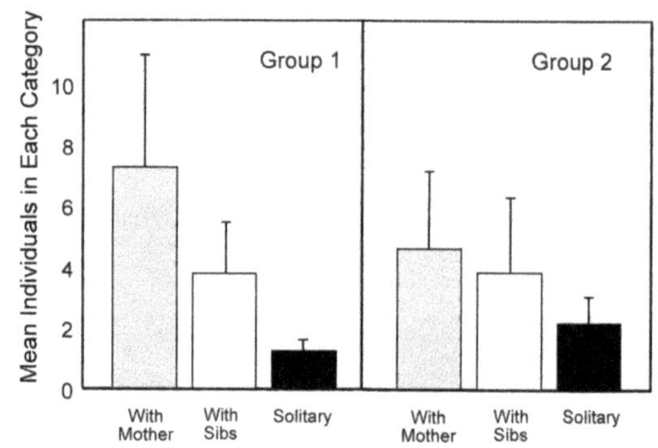

Mean number (+SD) of *Phrynus marginemaculatus* young in Groups 1 and 2 found in association with their mother, in sibling groups, or solitarily. Data from Group 1 (n = 14 observations) and Group 2 (n = 26 observations) were collected on an approximately weekly basis from age 1 month to 4 months.

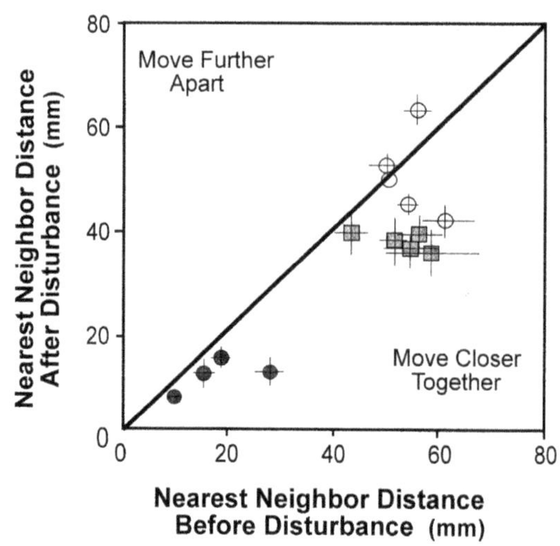

In response to disturbance, *Damon diadema* (Groups 5 and 8) aggregated together significantly more than expected. Data show the ratio between the mean nearest neighbor distances immediately before and four minutes after a disturbance. Circles represent replicates of Group 5 at age 1.5 months (< 30 mm) and at 8 months (> 40 mm). Squares represent Group 8 at age 2 months. Mean Standard Deviation for nearest neighbor distances before (vertical lines) and after (horizontal lines) are indicated in each datum.

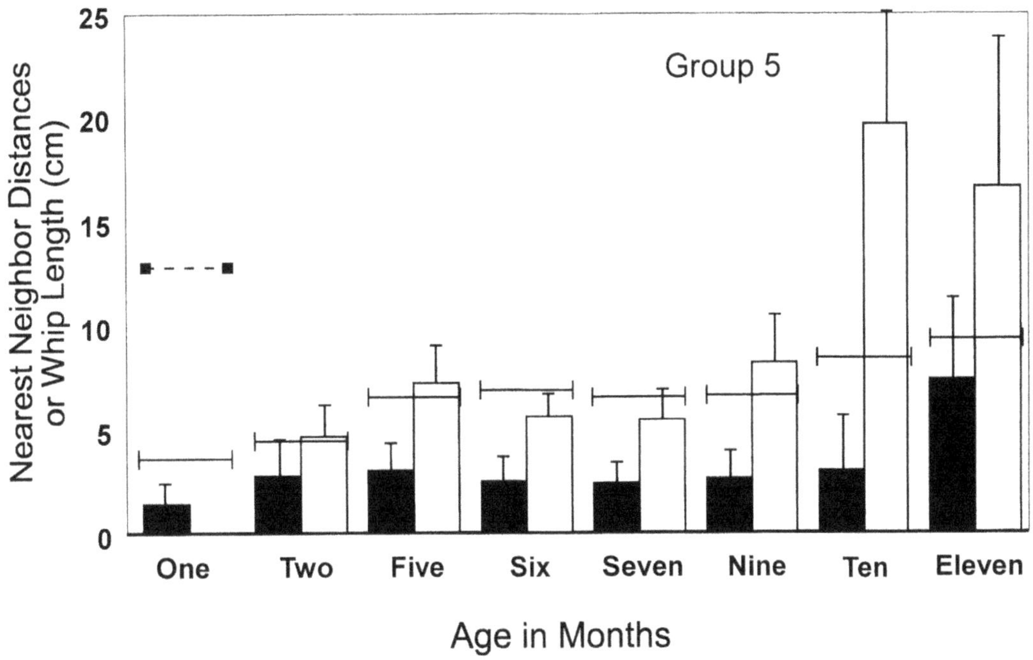

Nearest neighbor distances among *Damon diadema* in Group 5 relative to whip length for individuals in sibling groups (black bars) and solitary individuals (white) at different ages. Mean whip length of the young (I—I) and their mother (I– – –I) is indicated. Individuals in groups of siblings sit significantly closer together than necessary to make whip contact with one another.

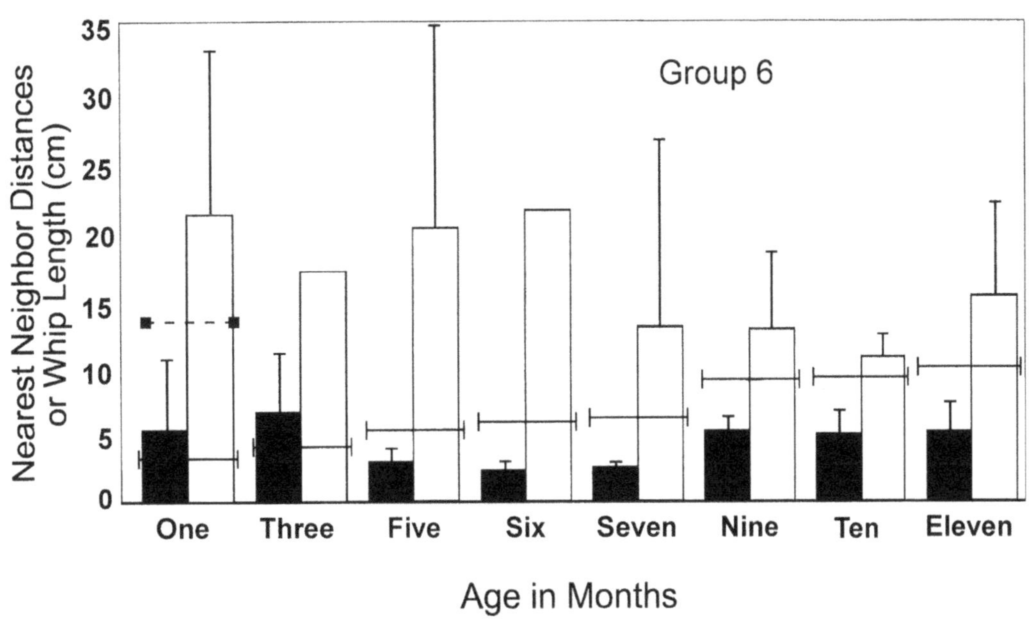

Nearest neighbor distances among *Damon diadema* in Group 6 relative to whip length for solitary individuals and individuals in sibling groups at different ages.

Some days this cork hollow is filled with specimens and other days none are present. *Paraphrynus mexicanus*.

Damon mother and old immatures. The clean-up-crew are "giant canyon" isopods, *Porcellio laevis*. © Ron Wagler

These *Phrynus marginemaculatus* under a piece of egg carton may have some communal tendencies but mostly chose this hide because it has preferred surface qualities and other available hides in the habitat were less desirable.

relative to familiarity in *D. diadema* for pairs unrelated to kinship. For introductions, it is important to remember if animals are separated for a few weeks or months they forget. When placed back together, serious aggression can occur even if the same individuals previously cohabited peacefully (McMonigle 2008). Reintroduction of a *Heterophrynus batesii* pair after many months, detailed in the reproduction chapter, included a somewhat tense introductory period. Nevertheless, introduction of specimens in the first few instars requires no introductory period since there is normally no aggression early in development.

As they grow older, amblypygids will cannibalize, but at such a slow rate it can be mistakenly attributed to other factors. Among my original *Damon diadema* sibling group I lost the first one after nine months and lost two more over the next two years. For a long time I assumed deaths were only the result of rogue crickets combined with bad timing (the communal tendencies having been exaggerated in husbandry circles). They had only been partly consumed, as consistent with crickets. Much of the available information claimed they did not bother each other and I witnessed a few molts where siblings seemed to almost watch over and protect each other. However, I chanced upon the last one to die (after the rest had reached maturity and mated). It had just molted—it was green—and half of it was sticking out of its sibling's mouth. Although cricket bite damage is often uniquely obvious, I had to wonder which of the previous deaths were caused by crickets. None had died prior to 5th instar but there was also less risk from crickets during these early instars due to less intricate caging and smaller prey (McMonigle 2005).

These 3rd and 4th instar *Heterophrynus batesii* are often observed together in groups of two to six, but individual specimens are not rare.

Aggression usually does not occur during the first five to six instars. In later instars, if the animal is not molting, aggression can also be in the form of fighting, where the loser is damaged or killed but not eaten. If a whip is broken off it is necessary to remove the animal before further damage is inflicted. Early instar whipspiders have incredible appetites and easily eat animals larger than their own mass, and yet cannibalism at this time simply does not happen. Even among late instar animals, cannibalism is commonly limited to molting specimens and is more common if animals are hungry or of vastly different sizes. Since molting animals cannot communicate with

These *Paraphrynus mexicanus* are intentionally clustering after a disturbance. Other acceptable hiding areas are available.

Paraphrynus and *Phrynus* cohabitation showing the chosen *Paraphrynus mexicanus* compared to a large female *Phrynus marginemaculatus* on one of the larger hides for size reference. Also depicts common proximity of cohabitating specimens.

the teneral whips. it likely occurs at this time for this reason. Still, many insects molt around the time whipspiders are most active, so teneral exoskeletons may trigger a significant feeding response. When large numbers of cockroaches are kept with whipspiders, freshly molted, white colored cockroaches are commonly seen being devoured. Nevertheless, cannibalized whipspiders are sometimes only partly consumed, which may suggest aggression over hunger. Cannibalism does not always occur during molts. I have observed cannibalism in late instar *Heterophrynus* unrelated to molts and a probable case of a third-grown *Paraphrynus* eaten by an adult. Paired adult *D. diadema* and *H. batesii* molt under the same hide without cannibalism.

The two filial generations of *Damon diadema* (including multiple broods from different pairs as well as some repeat broods) were kept together during the early instars and then separated at 5th-6th instar into larger cages with multiple hiding spots and reduced numbers of individuals, only six to eight per 10 gallon (38 L) terrarium. Over the next six months to a year some would be lost or maimed due to aggression. On my fourth generation of the same stock I left a brood in a 1-gallon (3.8 L) ice cream pail with a single cork bark slab and decided to wait for some signs of aggression before separating. They all reached maturity and molted once or twice after maturity, by specimen, without a single casualty. They pretty much had to hang from each other when molting. From this experiment, it would seem my previous attempts at trying to circumvent aggression helped cause it. Facultative sociality may be a common occurrence across whipspider genera for late instars and adults that are not given the space to set up territories are forced to keep in constant tactile communication. However, in this tight setup there was another unforeseen result that was not positive. Not a single spermatophore nor broodsac was produced by any of the twelve animals during two to three adult stages. What seemed to be a boon was a bust.

Siblings rarely stay away during the molting process. *Heterophrynus batesii.*

Social behavior generally must have some payoff or reward to the individual or group survival. The most common payoff is protection in various expressions, from the ability to overwhelm predators through distraction or sheer numbers, to creation of massive colonial shelters that protect from predators, desiccation, and flooding. Another common reward is the ability to catch and kill larger or more prey. Social behavior in regards to rearing young has

This mated couple often rests during the day in this position.

Facultative communal second stage adults, 4th generation
(5th were wild parents included) *Damon diadema.*

Phrynus barbadensis female chewed on by the male she had been living with following a molt (regenerated whip and partial leg), 2009.

the obvious advantage of increasing the survival of one's progeny. Social patterns in mating behavior can serve more or less certain functions since many creatures do not associate following copulation. In the Amblypygi, social behavior of mating pairs may increase the certainty of successful fertilization. Specific reproductive development likely means the male must stick around to ensure paternity. Success may be an issue since they mate multiple times over a week or two in captivity until fertility has been assured. They do not share food and likely offer no protection to each other. The mother's care must certainly protect progeny during early, vulnerable stages, while sibling social behavior may increase individual survival even if very minor protection is afforded only during the vulnerable molting period. Other social patterns are likely to give whipspiders the ability to share what can be a very scarce resource: protective daytime shelter. Whipspiders are among the very rare arachnids that can neither burrow nor construct shelters from silk or other materials. There is often plenty of surface area in most habitats for foraging, but extremely limited protective shelter for large species or for small species during dry periods. Consider facultative sociality (above), adult *Heterophrynus elaphus* shelter sharing (see species list), *Paraphrynus mexicanus* daytime shelter sharing in the wild (see species list), and dry season shelter sharing *P. marginemaculatus* (see species list). The species best known for abnormally intense intraspecific aggression, *P. barbadensis,* is also rare in that it has a number of physical features (and behavioral patterns) that allow it to spend the day on the surface of a tree rather than within deep cavities. In its case, foraging and shelter areas may be roughly equal. Lastly, solitary reports for some species may be misleading since the vast majority of reports are for specimens hunting at night, not encountered in daytime shelter.

Interactions should be monitored when animals are initially introduced or when young reared together in captivity grow large. Aggressive behavior often precedes death and dismemberment by days or weeks. Mating behavior is usually more ritualized and less violent, but can sometimes be difficult to discern from aggression. If uncertain and behavior gives the impression of aggressiveness or anger, it is best to rehouse animals immediately. (Fighting will usually grow worse at night.) Some individuals may be more aggressive, starving, near a molt, etc. If aggression is noted, a light smack to the side of the enclosure will stop aggression and the animals can be separated. Whipspiders are more attuned to outside stimuli than each other. Most other invertebrates have to be literally torn apart once they have begun a scuffle.

WHAT TO WATCH FOR: COMMON INTRASPECIFIC BEHAVIORS

(Compiled from behaviors of the listed *Damon, Heterophrynus, Paraphrynus,* and *Phrynus*)

Whip taps—These are generally slow and probing, though they do not always require direct contact with the other animal. Whips are generally very active in initial introductions, but are used now and then to confirm the presence (and position?) of cage mates.

Siblings nearby during a molt, *Damon diadema*.

Whip flicks—Generally only seen during initial introductions or mating, so it may relate to interest or aggression. Antenniform vibrations are seen as rapid jerking movements, not waves nor oscillation. Unlike whip taps where whips often move independently, this involves both whips (even if one is small or half missing) moving at the same time and approximately the same movement. The pattern seems to be somewhat different across species.

Palps open up—If movement is slow they may be measuring each other and there should be no attempt to touch palps. Often the palps open up in a warning gesture, but they do not contact the whips or palps of the other animal.

Grabbing or touching palps—Measured movements are of little concern but rapid movement and touching the other animal's palps or body is a sign of aggression and they should be immediately separated. (This may look similar to later stages of courtship in some species but should not be rapid.)

Palps angled—V-shaped posturing during mating or initial approaches.

Hand wiping—Palps are extended and the tarsi are wiped back and forth across each other rapidly. This is a sign of agitation commonly seen alongside other aggressive behavior.

Palps physically push object away—This occurs when animals feel threatened or annoyed and is more likely with prey items.

Chelicerae—If these are extended in reaction to another animal, murder is the goal. However, specimens often extend the chelicerae during grooming activities when sitting around near each other as they clean the palps, tarsi, and whips.

Body lifting—Long legs often raise up the body while scanning another whipspider, but also may lift up to get the upper hand in attack or rhythmically raised and lowered in mating courtship (*H. batesii*).

Body arching—Females may arch the body in an attempt to shield smaller animals even of different species.

Backwards retreat sideways—Generally slow and occurring when an undesirable animal enters a mating, scanning or hunting territory, and is warned by palp opening from the other. Since movement is often sideways this retreat is not fully backwards but rather backwards across the other's territory.

Backwards retreat full—Again slow and normally seen in back and forth movements of a pair during courtship.

Running sideways—Extremely rare in intraspecific reaction but is a sure sign of fright.

Clustering when disturbed—Early instars reared together generally cluster in groups under or inside objects if greatly disturbed. Adults and late instars kept communally often run in opposite directions (except for *Paraphrynus mexicanus* which notably cluster when disturbed).

Directional position—Pairs commonly face each other at rest while sibling groups are just as likely to face the same direction or towards the outside of center.

It is difficult to observe whipspiders without attributing human intent or feelings to their interactions. They seem agitated when they push each other with the palps or loving when they gently tap or rub each other or their offspring with the antenniform legs. The erratic tapping when a pair is introduced gives the strong impression of uncontrolled excitement, or fear. An established couple display little movement but they seem accustomed to each other's presence like an old couple in the park.

Cross-Species Introductions

Inspired by Rayor and Taylor's (2006) experiments keeping the mother of a species in with its offspring I recently decided to test cross-species interactions. I hoped interactions would be peaceful but was fully prepared to react immediately to any signs of overt aggression. I watched the introductions carefully for the first hour so I could intervene before attacks could escalate into mortal injury. This was not *Monster Bug Wars*—my primary interest was not seeing my animals slaughtered or damaged in any way.

The first involved the introduction of a small adult *Paraphrynus mexicanus* to a *Phrynus marginemaculatus* habitat. I did not choose the largest *P. mexicanus*, just in case the size gave it too much of an advantage since its legs are so much larger. The specimen chosen was a little smaller than the larger full-grown *P. marginemaculatus* in body mass. Originally there were four adult *P. marginemaculatus* in a 10-gallon (38 L) enclosure with four large and five small hides. The largest hide was a piece of 3" x 4" paper egg carton, which is compartmentalized and could be considered multiple hides in itself. The remaining hides were variously shaped pieces of tree bark that at most offered three surfaces each (front, back, and widened bottom). The large hides offered cavities from 1-2" (2.5-5 cm) in height. The small hides offered shallow cavities less than 1/5" (<5 mm) high, which are commonly employed by *P. marginemaculatus*, but never by *P. mexicanus*.

I observed no aggression in the first half hour. The relative placement of whipspider specimens in the hides was observed twice daily for one week. There were multiple hides and only five total animals, but only once did I find the *P. mexicanus* more than half an inch away from a *P. marginemaculatus*, alternating between two of the large wooden hides and the egg carton. I was surprised to find the *P. mexicanus* under

Behavioral repertoire involved in social and foraging interactions of immature *D. diadema*. Behaviors are divided into two categories: (a) those involving the whip and (b) those involving other parts of the body.

(b) Body postures, movements, and behaviors

Behavior Name	Category	Description
Slow scan	N/A	Specific details of movement may vary, but whip movements are slow and context indicates that the animal is relaxed and carrying out routine scans of the area nearby.
Inspect	Interaction	Angle subtended by whip motion is small; whip movement is focused in a specific area, but whip does not necessarily contact another object or animal.
Touch whip	Interaction	The focal individual makes contact with another amblypygids's whip.
Touch other amblypygid	Interaction	An individual touches its whip(s) to any part of another amblypygid's body, excluding the whips or palps.
Touch palps of other	Interaction	The focal individual touches its whip(s) to any part of the palps of another amblypygid or to the nearby region (chelicera).
Touch by other	Interaction	Focal individual is touched by the whip of another amblypygid.
Touch non-amblypygid	Interaction	An individual makes whip contact with a non-amblypygid object (e.g., a prey item).
Flinch whip	Interaction, Aggression	The whip is briefly and quickly jerked, usually backward. Most commonly seen after animal contacts another amblypygid or prey item.

(b) Body postures, movements, and behaviors

Behavior Name	Category	Description
Relaxed closed	N/A	Amblypygid's body is flat against the substrate, and palps are closed. Typical resting posture.
Relaxed open	N/A	Amblypygid's body is flat against the substrate, but palps are open. Contrasts with palp opening during aggression in that palps loosely drop open, often with the femur at $\sim 45^\circ$ hanging downwards. Occurs in relaxed contexts, as a typical resting posture.
Agonism low	Aggression	Palps are opened such that the angle between the femur and tibia is less than 90°. Right and left palps are usually in symmetrical positions. Palps typically open rapidly and are briefly held open before returning to the "relaxed closed" posture.
Agonism high	Aggression	Similar to Agonism low, but femoral/tibial angle is greater than 90° and typically held open for a significantly longer period. Indicative of a higher level of aggression.
Bat/swat palps	Aggression	One or both palps are opened and swung through the area in front of an amblypygid. The speed of this movement is generally not as fast as that seen when an individual is making a grab at another amblypygid or prey item. Possible context: locating another animal prior to making a grab at it.
Grab	Aggression	Both palps are moved simultaneously towards another animal. This movement is quick and clearly aggressive (possibly an attempt to remove the other animal's whip or leg). Movements are essentially the same as during prey capture, but context is interaction with another amblypygid.
Body elevated off substrate	N/A	Legs are straightened more than during typical resting posture, such that the amblypygid's body is no longer in contact with the substrate. May indicate heightened alertness; often seen in contexts of aggression or hunting.
Body pump	Aggression	Body is moved up and down as amblypygid bends and straightens its legs. Contexts are aggression and hunting.
Walk	N/A	Amblypygid walks across substrate, usually conducting whip scans as it moves.
Retreat	Aggression	Focal individual moves backwards, usually away from another amblypygid or prey item.
Erratic Run	Aggression	The animal moves across the bark at a speed faster than normal walking, periodically stopping and starting. Usually accompanied by quick movements of the whips in broad arcs. Animal often appears to be agitated, making efforts to quickly investigate its new surroundings.
Fast run	N/A	Usually seen in response to touching the base of an amblypygid's legs. The animal moves across or to the other side of the bark extremely quickly, apparently without stopping during the run to make whip inspections of its surroundings.
Orient to prey/prepare to attack prey	N/A	Palps are drawn back to open position—femoral/tibial angle usually is not more than 90°, but angle between palp and body is small. Both whips are usually in the area near the prey item, repeatedly and briefly touching the prey. The amblypygid's body is oriented facing the prey. This behavior is usually followed by grabbing at the prey.
Grab at prey	N/A	Both palps move synchronously from open position towards prey.

(b) Body postures, movements, and behaviors

Behavior Name	Category	Description
Capture prey	N/A	Prey is impaled on the metatarsal spikes, at the end of the palpal hands. If the first capture attempt is unsuccessful, an individual may make several additional grabs at the prey item.
Drop prey	N/A	Prey is released from palpal hands. May occur while prey is still alive, or after it has been killed; may be accidental or intentional.
Eat prey	N/A	Amblypygid holds prey in palpal hands during consumption, may move hands, and adjust position of prey. Chelicerae move up and down, masticating prey.
Groom	N/A	Includes several behaviors such as back and forth movements of the palps, drawing of the whips through the chelicerae, or chewing on the tip of the whips or legs with the chelicerae.

This table is used by permission of Linda Rayor.
From: Walsh, R. E., and L. S. Rayor. 2008. Kin discrimination in the amblypygid, *Damon diadema*. *Journal of Arachnology* 36: 336-343.

Greater density has been maintained in hopes of limiting developmental aggression in the F2 generation. 6th instars beginning to molt to (7th) 1st stage adults and not one instance of damage or cannibalism noted.

This mated pair (2013, female 2) rest during the day together as seen here, sometimes the palps touch, but they stay clear of each other during nighttime hunting. This followed the second introduction and the female began feeding again.

the egg carton twice since it had to crawl under the edge—use of shallow hides or crawling under objects is not a behavior observed in the communal *P. mexicanus* habitat. With thirteen of fourteen observations showing touching animals, this was more than was seen with the *P. marginemaculatus* specimens in the same habitat towards each other. Results were consistent with *P. mexicanus* behavior, as this species seems more social. (For example, there is a coconut hide in the *P. mexicanus* habitat that is only occupied two days a week, but when it is, there are always two or three animals, never one.) A few nighttime observations were also made and no aggression was noted, just normal solitary scanning and hunting behavior. After the first week, twelve additional *P. marginemaculatus* were added to see if interaction would change, since with this many it would be difficult for any specimens to find shelter not occupied by another animal, despite the large number of shallow hides. After seven more days, no new behaviors were noted and the *P. mexicanus* was removed.

The second cross-species test involved a large adult *P. marginemaculatus* female introduced to a mated pair of *Heterophrynus batesii*. I expected the *Phrynus* to run away rapidly and hide in a crack. On initial introduction there was immediate whip scanning by all three animals and as the male *Heterophrynus* showed interest, the *Phrynus* hid under the female. The *Phrynus* came back out after thirty seconds and touched whips with both *Heterophrynus* for twenty minutes. All movement was slow and measured. When the enclosure was bumped the *Phrynus* ran back under the female. A minute later it came out from under her and continued to touch whips with both *H. batesii* for another twenty minutes and then it slowly left and hid under a log on the other side of the enclosure. I had planned to remove it the next day but it found a hole in a piece of wood and it took another six days before it was found roaming and could be removed from the terrarium safely. (If I had broken the wood apart to extract the animal, it could have easily been damaged or killed.)

Heterophrynus batesii are aggressive predators and would have taken out a similarly sized cricket in less than a minute.

I was surprised at the amount of whip touching and lack of any overt aggressive or excited behavior. (I have observed some heavy aggression when introducing pairs or multiple adults of the same species.) The above introductions were not designed to prove a specific behavioral response pattern and I would not suggest keeping different amblypygid species together for months or years. However, the responses do suggest whipspider aggression may be rooted almost entirely in courtship and territorialism, which has been recorded most notably for captive *Damon diadema* (McMonigle 2005, Rayor 2006 and 2007) and *H. batesii* (McMonigle 2011). The responses appeared to be based on tactile input, though olfactory cues cannot be ruled out. Near-sound communication appeared entirely lacking as no rapid, antenniform leg vibrations were observed.

Social Verbiage

Discussion of social definitions is placed at the end of the chapter because the various terms used almost never adequately describe the behaviors exhibited or the changes that take place during different stages in development. The specific definitions of related terms are debated by various authors. The complex societies found among certain species of eusocial hymenopterans and termites are often considered the only true examples of sociality in invertebrates. Yet termite and ant cannibalism is common and well documented and not just in response to adverse conditions. Other creatures that show some sort of social pattern, ranging from limited motherly care of eggs to cooperative hunting and retreat building, fall somewhere below the eusocial spectrum under various subsocial terms including communal and quasisocial. The terms are like round holes into which every square peg must be pounded. Of

Cross introduction start, *P. marginemaculatus* and *H. batesii*. See following pages for additional notes.

course the incredibly complex behaviors and variable social interactions displayed by the different whipspider species, their expression under varying conditions, and how such behaviors relate to those of other social creatures could not be expected to fit into a lifetime of study, let alone be perfectly encompassed by a word or two. Rayor and Taylor (2006) suggest *D. diadema* and *P. marginemaculatus* (and considering present experiences with *Heterophrynus* and *Phrynichus*, likely the majority of species in their respective families) be labeled "prolonged subsocial." This is meant to describe complex, long-term associations between mother and young or among sibling groups that may last as long as sexual maturity, but not until there are multiple breeding adults in the group.

The *Phrynus marginemaculatus* is hiding from the interested male who seems to be showing more interest than the female. The female *Heterophrynus batesii* can be seen arching her abdomen up in a protective stance.

Cross intro, after thirty minutes interest is visibly waning, specimens have stopped scanning or posturing.

4
HABITATS

AMBLYPYGIDS, though frightening in appearance, are harmless, timid creatures that hide away the day in small caverns at the base of trees, old logs, caves, rock faces, and animal burrows. A few species are known to inhabit the hollows of termite or ant nests, but not one species is known to construct or excavate its own shelter. Despite the popular notion, for many species, the vast majority or the entire population is found nowhere near a cave. The so-called "cave spiders" eaten on a popular television show were not only not spiders, but also were not collected from caves. They were collected from the buttresses of large trees. Most shelters, such as the cavern below an old log, inside a rotten tree, under a stone, or within a nest, do not qualify as caves. Even cracks within a large rock face are not considered caves by human standards. No dead or live specimens presently available to hobbyists in North America originated in a cave. Specific first-hand details of habitats are covered individually in the species list. One reason for featuring the Asian *Charon grayi* is this species' habitat photographs are among the few I could find with the barest documentation that included a cave. (I could photograph a local cave as is often done to illustrate theoretical habitat, but all habitats shown in this chapter were chosen for their authenticity and whipspiders are not found in my state.) Furthermore, even *C. grayi* and the Mexican *Paraphrynus* sp. seen here are encountered hunting outside the cave and likely only use it as a daytime retreat, so a cave would only be one portion of the required habitat.

Whipspiders, particularly the largest species, rely on massive, old trees for habitat. Hebets (2002) studied wild *Phrynus parvulus* on trees in Costa Rica and found they select microhabitats based on the tree surface, moss cover, and presence of buttressing. Large trees in the preserve could host a number of individuals, but large trees were also the most common. Dias and Machado (2006) studied one of the biggest, *Heterophrynus longicornis*, in its natural habitat and concluded undisturbed forest and very old trees are preferred habitats. They found that selective logging and forest fragmentation adversely affect populations and *H. longicornis* may already be locally extinct in various forest fragments. Weygoldt (2002) also remarks that members of the genus *Heterophrynus* are eliminated from areas when the forests are cut down (as opposed to tiny Amazonian *Charinus* and small *Phrynus* that may make a living in more varied habitats and can be found under stones or on young trees).

Puebla cave *Paraphrynus* sp. © Eddy Hijmensen

Puebla cave species. Close-up on eyes, the median eyes are reduced. © Eddy Hijmensen

This *A. coronatus* pair is on a tree trunk in the habitat. © Eddy Hijmensen

As forests are eradicated only those species or populations limited to real caves and rock faces may be able to survive, but even these could be adversely affected by massive shifts in the surrounding habitat and pesticides used on prey that make their way up the food chain. The whips and impressive repertoire of tactile and chemical sensors gives this order a decided advantage in cave habitats. Some species are true

The stone in the picture is limestone caliche cliff face and this Texas *Phrynus operculatus* is about three feet (1 m) off the ground. © Tripp Davenport

"Who told you to bring me flowers?" Amblypygids are not found on flowers in nature and may even find the smells or textures repulsive.

Puebla cave habitat, Mexico. © Eddy Hijmensen

cave dwellers, particularly in the Charinidae and Charontidae, while others prefer caves (or cave-like man-made structures like bunkers, wells, or tunnels) even when the surrounding habitat is undisturbed (Schwarz pers. comm.). Of course, even the most isolated cave ecosystems have been destroyed by human interference in the surrounding habitat. Man may be the only serious threat to the existence of whip-spiders, but may also be their primary means of survival. Captive propagation, especially where habitats are most vulnerable or no longer exist, may prove useful for the conservation of amblypygids. Raising awareness may help change opinions and raise sympathy for Amblypygi, lessening the chances of people killing them out of fear or hate. Improving awareness may prove impossible, but the details provided here can prevent those who do care from accidentally killing their charges through use of improper habitats.

An argument could be made that captive-reared stocks would reduce collection pressures in other countries, especially for dead stock. However, if native people cannot make a living by sustainable harvesting of local fauna and flora, the alternative is not uncommonly to burn down the forest so they can plant crops. Arbitrary biological idealism usually does not trump personal survival for most humans, or at the very least for keeping one's own children from starving. Legislation is often limited to banning collection or captive maintenance rather than saving habitat because habitat preservation requires resources, while bans are quick, require no upfront expenditure, and are politically correct. This unfortunately results in habitat not being saved and the few persons who might care or notice being silenced. There are many species of fish, including various killifish, Mexican livebearers, and pupfish, as well as two cave cockroaches, whose habitats have been utterly decimated in nature through the mining, drilling, and agricultural economic activities of man, which now only survive in captivity. Knowledge of and interest in captive husbandry may be the last hope for some species of whipspider. The large species are often easier to propagate, are the most enticing, and are also the most at risk of being eradicated in

These West African *Simandoa conserfariam* are believed to be extinct in the wild due to habitat destruction of a cave system resulting from bauxite mining. They continue on in captivity.

Chamela reserve in Jalisco, Mexico, (Jalisco dry forest) habitat *Acanthophrynus*. © Eddy Hijmensen

Chamela reserve in Jalisco, Mexico, (Jalisco dry forest) habitat for *Acanthophrynus*. *Phrynus* are found in the same habitat, under rocks and in burrows along the roadside. © Eddy Hijmensen

Phrynus sp. habitat in Colima, Mexico. © Eddy Hijmensen

Phrynus sp. (Oaxaca, Mexico) in habitat under rock. If one did not have the colorful native rock as a reference, the bright colors could be mistaken for warning rather than cryptic purposes. © Eddy Hijmensen

Charon grayi and *Sarax curioi* habitat, Philippines. © Christian Schwarz

Charon grayi habitat, Philippines. © Christian Schwarz

Charon grayi habitat, Philippines. © Christian Schwarz

Charon grayi, on tree in secondary forest, Panay Island, Philippines. © Christian Schwarz

nature since they require more specific habitat. It would not take incredible effort, nor a large number of people, to maintain a species indefinitely, as whipspiders can be easy to propagate when care requirements are understood.

Before explaining the specifics of captive habitats, the natural habitats of various species are shown here. These include close-ups of wild animals in the possible, preferred microhabitat and wide-angle habitat images. (All photos are actual whipspider habitats where specimens were found.) Complex artificial environments may be created in hopes of copying nature, but the specifics of the microhabitat an individual animal is found are not likely to be understood even with a limitless series of amazing photographs. Furthermore, a scientific approach to husbandry means that variables must be limited to test their use. A beautiful terrarium which may seem to have "all the comforts of home" may in fact provide multiple dangers and lack simple, necessary requirements. A pretty cage is of no value if it does not offer suitable molting surfaces or if it contains features that lead to drowning, dismemberment, or escape. One likely reason many species are limited in nature to large trees and rock surfaces is, unlike many insects, they cannot or will not molt off of most plant stems or leaves. Some keepers include small fountains and living plants in large, intricate setups. Colorful and intricate vivarium setups may be attractive, but make it far more difficult to monitor feeding, excess prey, aggression, etc., and may take away from, rather than add to, the lives of the animals. Is the goal a pretty cage, maximum survival, or a little of both? Will an adequately considered habitat with beautiful floral decorations add anything to the life of animals that probably cannot see color and might even be repulsed by the smell of flowers? They are certainly not

Inhabitant close up in habitat below. Prime *P. marginemaculatus* habitat in Lee County, Florida. © Paul Bollinger

Phrynus marginemaculatus habitat in Lee County, Florida. © Paul Bollinger

Paraphrynus raptator habitat. © Floyd & Bollinger

Paraphrynus raptator habitat. © Floyd & Bollinger

One of the Lapuna trees where *H. batesii* can be found in Las Palmas area, state of Lareto in Peru. Far left, Americo Martinez, sitting Paul Bollinger, and friends. © Sam Floyd

Paraphrynus mexicanus habitat, Pinal County Arizona. © Peter Clausen

attracted to flowers. Consider the explanation of individual requirements discussed and then back track out to add accessories or decorations. Do not start with the cage in mind, but begin with the requirements of the animal. Of course the text here can only offer specifics of what has proven functional and attempt to explain things that have been known to cause problems. There will not be much discussion of untested husbandry ideas. (You may find a greased jello mold makes the best molting surface or that your animals like to take swims, but I have not tested either).

Enclosure Considerations

In captivity, habitats are usually made from plastic storage boxes, deli-cups, or the glass boxes normally used as aquariums. My strong preference is towards glass terraria since these allow for easy viewing, are readily acquired, and high humidity can

Rock face in Tucson, AZ, where *Paraphrynus mexicanus* is sometimes found. © Ryan Eide

be maintained. (Plastic boxes often have large gaps around the lid and restricted viewing.) For small enclosures, deli-cups can be nearly as clear and are useful to separate large numbers of small animals.

The two main habitat types used in captive rearing are large communal and small solitary. (A mid-sized cage may be more reasonable for housing single pairs.) Large communal habitats are commonly used for groups of animals since most will cohabitate, and a large solitary enclosure may be too much space to dedicate to every animal in the case of reproduction. The large habitat is easily made from the common 10-gallon (38 L) glass boxes sold as aquariums. The bottom is lined with 1″ (25 mm) of substrate such as potting soil, coco fiber, or peat. Leaf pieces or small chunks of wood among the substrate should be avoided because such items protect prey. It is also a good idea to firmly compact the substrate by hand. Finally, large slabs of bark or wood are set at an angle to provide areas for hunting, mating, and molting.

The small, solitary cage is made from a 16 oz (0.5 L) or 32 oz (1 L) deli cup with three modifications. This enclosure has been successfully used to rear hundreds of animals with an almost perfect molt record, though larger cages should be constructed for postultimate molts of *D. diadema* and larger species. A 1-2″ (25-51 mm) square hole is cut in the side and covered with screen for ventilation. Half a dozen thumbtack holes may be used for humid species. A circle of metal window screen is cut to fit inside the lid where it will be glued to serve as a molting surface. If this is not glued in to cover the bottom of the lid or if the lid is not readily accessed by climbing on a high cage decoration, the enclosure must be far larger to account for the loss of this important molting surface. The diameter of the molting surface must be at least as large as the specimen's walking leg span. A 3/4″ (19 mm) wide, screen "ladder" is cut to run from the bottom corner to the opposite top corner. Lastly, 1/2″ (12 mm) of soil is placed on the bottom.

32 oz solitary cage used successfully to rear large numbers of *Damon diadema.*

Both caging options have good and bad qualities. In large communal cages the immatures almost always mature faster

10-gallon (38 L) *H. batesii* communal brood cage used to rear F1 & F2 generations.

Phrynus marginemaculatus on egg carton, temporary overcrowding causes no harm but this species prefers to spread out if given the chance and should not be kept in high density for more than a few months or losses may be seen.

and are easier to feed. Some specimens are always hungry and so a group can protect its members from aggressive prey only in as much as they consume it. (There is no apparent deliberate protection but in captivity the end result is the same.) Despite the benefits, as they grow close to maturity, intraspecific aggression can become severe. It is generally easier to maintain humidity in the large habitats, but closer attention to feeding can be required to limit aggression in older specimens. Individual cages eliminate cannibalism and often take up less space. (Dozens of solitary cages for larger *Damon* immatures can be kept in the same space as a 10-gallon (38 L) that should not house more than three or four large *Damon*.) While some breeders have had great success with communal cages, luck may not be on your side. These are predatory animals and the only sure way to ensure cannibalism does not occur in later instars and adults is to keep them apart. However, while highly social ants cannibalize in the absence of food, starvation has never caused cannibalism in early instar amblypygids.

Cage decorations that provide molting surfaces are extremely important for the whipspider habitat. If lack of an appropriate molting surface forces them to molt on the ground, they will lose whips, become deformed, or die. These are arboreal creatures that hang out high on tree bark and rock faces in nature. Unlike most pet arachnids they must molt while hanging upside-down from a grippable surface. They are not capable of climbing or hanging from smooth glass or plastic surfaces during a molt. Even the tiny species, not kept in North America, that have glass climbing ability would probably fall if forced to molt from a smooth surface. For this reason, molting surfaces should be chosen that are porous and allow the animal to grip easily without risk of falling during the molt. Natural rock faces are often variably rough or have moss growing on spots, so there is no reason to believe amblypygids will molt successfully from a differently porous rock in captivity. Also, the two commonly available

Molting height does not need to be nearly as great as commonly suggested since whips are extracted by palps.

The whip is almost fully extracted. (Note all weight is held by tail.)

Solitary caging for *Phrynus barbadensis*. 5.5″ (14 cm) tall x 6″ (15 cm) diameter.

5-gallon (19 L) habitat for *P. marginemaculatus*.

species are normally found on trees or logs. They molt best hanging upside-down from a horizontal screen or angled cork bark, but also do well with vertical molting surfaces on bark. If they fall while molting they may twist or bend the pedipalps, and death will follow in a day or two.

The height of molting surfaces should be no less than the span of the walking legs. Somc hobbyists prefer incredibly high molting surfaces, citing the spectacularly long antenniform legs. However, the long front legs are carefully extracted from the exoskeleton by the pedipalps during the molt so extra distance is unused. Even if an animal does not die from a bad molt caused by too low of a molting surface, it will lose the whips or have bent appendages. Overly bent legs or whips will be autotomized even if initially freed from the exuvium. When both whips are missing, animals tend to grow slowly and can have a more difficult time feeding.

Cork bark slabs are probably the most commonly used cage decoration. The material is lightweight, resists mold, and is easily acquired. Cork half logs are a favorite hangout and are usually best employed standing on end. I have used a half log made from coniferous wood (cedar) for years with great results but it was aged before I acquired it. This cedar half log shows up in quite a few photographs, including the cover. Coconut huts are included in some cages but are never a favored retreat and only useful for rather small whip-spiders. Cork bark, half logs, and coconut huts are commonly available at most pet shops or at reptile shows. I prefer natural looking materials like cork bark, though it can be expensive or not available in the size or shape needed. I have used large wood and bark slabs from willow, beech, and oak collected from outdoors. Extremely thin bark or pieces with many holes are often refused by the animals. Low cost, rough surfaced (not the expensive, smooth type) plywood ought to work well and can be cut to any size (though it will fall flat more easily and possibly cause damage). I have accidentally dropped cork and bark slabs on a number of animals over the years and have not yet maimed or killed one, but these decorations are not very heavy or flat.

Phrynichus cf. *phipsoni*. © Dan Wulf

Two specimens of *Phrynichus* cf. *phipsoni* were found under these large boulders (> 60lbs/ 30 kg) on a hill slope in the periphery of a dense forest, approximately 100 meters from a river. The location is a few miles southwest of Vellore, Tamil Nadu, India. © Dan Wulf

Rayor (pers. comm.) reports losing a number when cages were moved and bark fell on them. Wood, paper, or foam with a smooth surface may seem to work well for small animals, but should be avoided for large animals. The risk of death and deformity during a molt outweighs any benefits as the claws outgrow the pore size. Peat or coconut fiber can be glued to the surfaces of these smooth materials, or to customized structures made using expansion foam. Rocks such as slate and sandstone have been used but can offer disastrous molt results, eventually wear out claws, and are very heavy so they will almost certainly damage or kill any whipspider they fall on.

Metal window screen surfaces result in high molt success and animals often molt from the lid if it is easily reached. Screen or plastic ladders allow specimens to reach the lid in small cages. However, I offered sheets of metal window screen to provide alternative molting surfaces in large cages for years and never caught a single whipspider molting or hanging out on this material. The lid of the cage is normally covered, so it may register as a rough, solid surface when screen pieces do not. Small *Damon* and adult *Phrynus* are fond of paper egg carton.

Narrow cracks for hiding are not necessary. If tight-fitting junctures of bark or rocks are provided, some species may hide in them during the day, especially *P. marginemaculatus* and small *Damon*. Slightly curved pieces of bark can be set on the substrate for *P. marginemaculatus* if desired. Most large animals prefer to stand or rest flat against a large surface facing an open area below. In the case of *Heterophrynus* and *Paraphrynus*, these will refuse to hide in a crack during the day unless sufficiently disturbed.

Avoidance of hiding spots for prey should be a major consideration when setting up a cage. If there are too many shelters all the food being added may go to waste and lead to starvation or present a deadly molting hazard. Substrate should be relatively flat and fine, with no large chunks. Substrate is there only to retain moisture and maintain humidity in the cage. Leaf pieces are not a good idea. The substrate, decor, and type of hide structures chosen will prevent the use of certain prey types if the prey can stay hidden under or within. Wood or bark pieces with large or deep cavities should be avoided even with highly active prey that rarely tries to hide. Commercially available cork bark can be the worst culprit since pieces are often riddled with large tunnels that could easily shelter volumes of prey. Decorative wood and cork pieces should be chosen with care. If choices are limited, holes can be filled with hot glue or expanding foam according to size.

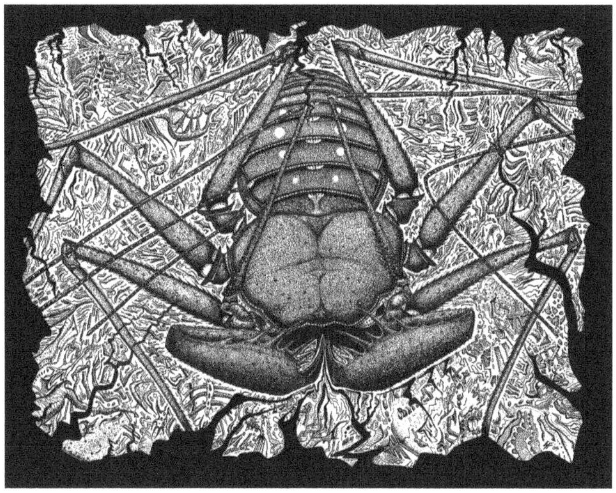

Phrynus art by Sam Floyd

Whipspiders can be harmed by too high and too low of temperatures. A temperature from 70°-75° Fahrenheit (21°-24° Celsius) is best for the listed species though 66°-80° F (19°-27° C) may be acceptable. The upper and lower ranges of temperature sensitivity are likely to be different for each species and a proper test would require the death of many animals which is both unrealistic and unacceptable. (Also it would only test upper and lower limits of survival rather than those best suited for long-term husbandry.) I have experienced increased problems with gestation and molting when specimens are kept at or above 80° F (27° C), while temperatures higher than 86° F (30° C) are known to be dangerous over long periods (Cernicka 2003). On the low range *Damon diadema* are known to die in

Abnormally colorful male *Heterophrynus longicornis* from Pará, Brazil. Five males, two females, and three immatures were found individually in cement pipe boxes near forested areas. When a pair was placed together in captivity they mated, and then the female ate the male. © César Augusto Chaves Favacho

shipping if the temperatures inside the shipping carton dips below 40° F (4° C) for too long, while *Phrynus marginemaculatus* can accept near freezing for short periods (pers. obs.). When subjected to these temperatures *Phrynus* will not move for at least half an hour after being warmed up. They can be laid on their backs with no movement. Nevertheless, the tarsi still grip surfaces, so specimens still cling to paper towel or other rough surface even when there is no apparent movement and the animal seems stiff and dead. Since temperatures in Florida dip very low on rare occasion, this may be a temperate adaptation that allows them to keep from falling from perches and being exposed to predators. This ability to withstand cold may not be found in its tropical Caribbean form.

The locations of listed species are given as much as possible, which means seasonal rainfall and temperature data could be acquired. However, at this point in time there is no evidence that temperature extremes or seasonal variations found in amblypygid habitats play any role in proper husbandry. The microhabitat is again a question since we do not necessarily know how desert animals avoid desiccation or where the animals hide during the heat of the day, let alone what those temperatures are. Fortunately, species kept so far seem rather content around the temperature humans are, and seasonal humidity extremes are not needed to trigger reproduction.

Small and immature amblypygids are particularly vulnerable to low humidity. Large specimens may survive a lack of moisture in captive habitats for months, but if they happen to molt during a dry period it can prove fatal. *Damon* and *Phrynichus* are able to survive rather dry captive conditions, but even they have trouble with low moisture for long periods. In order to regulate humidity within the enclosure a water dish may be provided, or an area of the enclosure such as a side or corner may be occasionally watered to be kept constantly damp. Watering frequency will be determined by temperature, airflow, and humidity, but should only be necessary once every week or two. If watering is needed more often because damp areas become bone dry quickly, ventilation should be reduced. If the enclosure is well-vented and ambient humidity is low, the decorations would need to be sprayed down lightly every day—if it dries rapidly, desiccation could still be possible. In northern areas the ambient humidity in winter can go below 5%, which can still dry out cages that have only a small vent. Airflow increased by a nearby vent or fan will increase the speed of drying. If exceptionally dry and well-vented, one forgotten watering could result in complete desiccation of small animals.

In recent years I have kept all the Phrynidae, including the two consecutive generations of *Heterophrynus batesii*, in terraria with solid glass lids to maintain humidity. *Damon* can also be kept this way, but I just dampen a small corner of the substrate and push the lid over to expose a 1/16" (2 mm) gap along one side. Keep in mind that unscreened ventilation, such as a gap

Mature male feeding, *Heterophrynus longicornis* from Pará, Brazil. © César Augusto Chaves Favacho

Phrynus sp. found in dry scrub habitat in Guerrero, Mexico. © Eddy Hijmensen

Puebla cave *Paraphrynus* sp., immature coloration. © Eddy Hijmensen

White micropod, *Trichorhina tomentosa*, SEM.

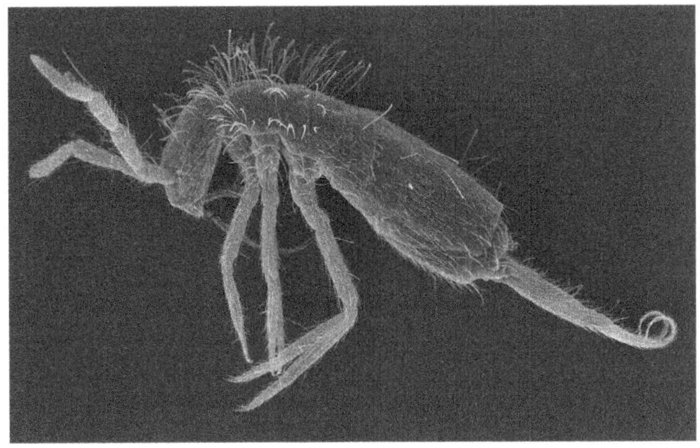

"Giant" springtail with tail sprung, SEM.

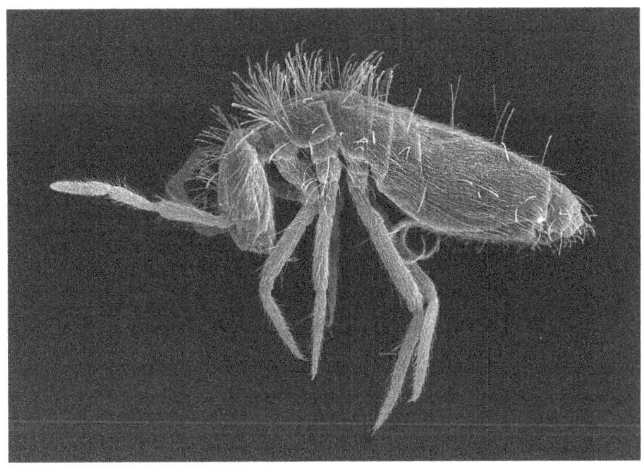

"Giant" springtail SEM.

along the glass, cannot be used if small glass-climbing or flying prey are used. In other species, half to three-quarters of the substrate is thoroughly moistened using a spray bottle. The solid glass lids I use are simply the bottom of old tanks; sides are a few inches longer than necessary. I find cracked tanks in the trash sometimes, remove the silicon with a razor blade and cover the exposed edges with duct tape. There is only a very tiny gap between the terrarium frame and glass but it provides plenty of ventilation. Limited ventilation has made rearing *P. marginemaculatus* immatures much easier, since earlier caging required daily checks when ambient humidity would dip very low during winter heating and summer air conditioning. A forgotten few days could easily wipe out an entire brood. Since I developed this method I have not tried to keep other species with higher levels of ventilation since it has not proven necessary and can of course be very dangerous.

Habitats should be kept clear of rotting prey items that may poison the air or cause proliferation of mold, as well as arthropods that might be harmful. Isopods or "giant" springtails are often employed as clean-up-crews since amblypygids are messy eaters and cage substrate is usually not allowed to become dry enough to kill them. For years I have used orange form *Porcellio* sp., but recently have been using the tiny *Trichorhina tomentosa*. Too few isopods will not consume materials quickly enough that could lead to mite or mold growth, though the exact number is determined as much by enclosure size and inhabitant numbers as it is by prey item size, type, and feeding frequency. After a year or two the substrate can seem to be composed more of isopods than dirt and clean up crew populations should be reduced. In practice, isopods are harmless even in plague proportions, because whipspiders do not molt on the ground, but isopods chew up wooden cage decor and consume dead, uneaten prey (a cricket the keeper thought was eaten by a whipspider may have only died). Isopods cannot prevent all types of mold growth. Excessive, cottony mold growth can make the habitat unsuitable. Small mushrooms or flat, two-dimensional patches of mold are of no concern, but wood with large patches of fuzzy or cottony mold should be removed and discarded. Massive fungal growths can restrict the animal's movements and incur excessive grooming. If mold is cleaned off the wood it often grows back in the same spot. Wood should not be heated or cooked for sterilization because heating can rupture the cell walls and result in excessive mold growth. Cork or wood slabs should be allowed to air dry thoroughly indoors for a few weeks to a few months to ensure large pests such as centipedes are not added to the habitat. Even tiny predators introduced from outdoors may present health hazards. Dryness and ventilation inhibit mold growth and prevent proliferation of many small arthropods, but extreme dryness is usually more harmful to the whipspider.

White micropod, *Tichorhina tomentosa.*

This *Phrynichus* cf. *phipsoni* from Tamil Nadu, India, reminds me of the giant ground beetles and domino roaches, also from India, that are black with white spots. © Dan Wulf

Habitat Recap

Amblypygids do not molt on the bottom of their enclosures. Rough, high molting surfaces are a necessity. Flat pieces of wood or bark should be placed at a high angle to provide adequate molting surfaces. Humidity and nutrition have a less immediate effect on molting success.

Communal housing makes it easier to feed without excessive monitoring. It is more practical to provide larger enclosures with better choice of molting surfaces. Immatures do well together for the first four to six instars, but they do not get along well in the "'teenage years" and should be separated. Most will again cohabitate safely in pairs or harems when mature or nearly mature.

Maintenance of high humidity and limited ventilation is acceptable and useful as long as some precautions are taken. Excess prey or decomposing and excessively moldy items should be avoided, and a clean-up-crew employed.

5
HEALTH AND REGENERATION

OVERALL THIS GROUP is quite hardy and with care it is very likely every single immature from a brood can be reared through multiple instars with no losses. This statement might be made for any creature, but it is true in reality, not just in theory, for most people who try breeding whipspiders. Unexplained die-off or defects are almost unheard of. The usual causes of death are not really health issues but use of dangerous or nutritionally deficient prey, unsupervised cohabitation, or defective caging. Still, there are a few things to consider in keeping them alive and healthy.

Whipspiders are very skittish and commonly jump or drop off surfaces if they are handled. They rarely crack or display external damage if dropped on a hard surface, but will die if dropped within a week of the molt, before or after. The long-legged species are more likely to jump, but even the clingy, two-dimensional *Phrynus* will jump every so often. Recently molted animals should never be handled for this reason. Knowing when an animal is a week prior to molting is not as easy to tell as after. Harrell (2006) says specimens he used in educational displays died earlier than those left at home, but the only *D. diadema* I know of that lived five years as a mature adult was brought to more than a hundred educational displays over that period. Although they are a joy to handle and pose no risk to the keeper, handling may pose great risks to them. The most docile can still respond quickly and sometimes you cannot be sure if you blinked or if the specimen moved to the other side of the cage, piece of bark, or back of your arm faster than your open eyes were able to register. None of the specimens held for genital operculum photographs were harmed in any way. Except for the rare 2nd instars with overactive autotomy and the freshly molted, careful handling and grabbing of specimens by an adult or mature teen poses no threat to their health. Of course the body must be grabbed, never the legs, as even minor twisting of a whip or leg may cause the damaged appendage to be autotomized a few days later. Handling should take place in an open, uncluttered area to prevent easy escape. Certainly, these are skittish, somewhat fragile animals—allowing a child or immature teen to handle one would be cruel.

Since they normally stay on cage decorations away from the substrate, mold and mites would have to be unbelievably bad to have any effect on health. Some white, cottony molds that grow on dead prey in damp spider habitats can kill their inhabitants

Entomophagus fungus feeding on a mygalomorph spider.

This flat, white mold growth on the substrate is not harmful (*H. batesii* spermatophore also pictured).

Grain mite outbreaks are able to harm whipspiders. These are pre-hypopus stage, but outbreaks are not always easily seen beforehand.

over time when the mold is in constant contact with the exoskeleton. This mold likely contains enzymes that break down chitin as it internally discolors the exoskeleton, but damage may be the result of other adverse properties. I have only seen pest mold with small Mygalomorph spiders a few times and if it does not take up much of the habitat or is removed within a few weeks, the animals are unharmed. If mold has grown into the specimen's exoskeleton, as evidenced by subsurface discoloration, special attention may keep the spider alive, but it will usually die before the next molt. This same mold is likely to affect amblypygids in the same way. Some of the mushroom hyphae and molds that feed on rotting wood can look somewhat similar but are harmless.

Rayor (2006) reports losing four entire broods of *P. marginemaculatus* in later development to mite infestations of the type normally associated with rotting crickets. I have not yet seen grain mites on wild-caught or captive whipspiders, but I do use isopods as a clean-up-crew to eliminate buildup of rotting prey refuse in the cages. Mite problems are, however, very common, and difficult to avoid with some other types of invertebrates. Grain mites are found almost everywhere. This group of mites are so named because they are common pests of stored product, but they also feed on various rotting plant materials and dead insects. The most commonly encountered is probably *Acarus siro* as it infests mealworm media, fruit fly media, and cricket cultures as much as arachnid and centipede enclosures. Full-grown reproductives are much smaller than a grain of salt and seldom are much bigger than the period at the

This is the late instar female *Damon diadema* who had a palp removed by a cage mate, two molts and a few years before she died.

This old male *Damon diadema* is already dead following a bad molt to 4th adult stage. Previous molts had been successful in the same cage under the same conditions.

end of this sentence. The hypopus (dispersal) stage follows an outbreak as food runs out. The mites molt into suction cup-like forms that tightly adhere to other invertebrate exoskeletons and can block breathing holes and lead to death if outbreaks are severe enough. If large numbers of tiny white mites are seen, the substrate should be replaced and the cage cleaned. This is most helpful when done before mites have begun molting into the hypopus stage. The final step, beyond employment of clean-up-crews, is to remove dead prey remains and not overfeed. Leaf litter (made from fresh leaves), dog food, and fruit can also lead to mite outbreaks, but these should not be in the amblypygid enclosure.

This *H. batesii* was found molting to 6th instar but the exuvium had somehow become unseated and the entire animal was hanging from one leg. Ten minutes were given to see if the animal could free itself but the molt was not progressing. Careful manipulation of the exuvium (not of the terribly fragile new exoskeleton) allowed it to start pulling its legs free as seen here but a palp was still stuck. Following an hour of careful manipulations it was able to free itself fully but the appendages were imperfect. The whips were severely warped because they started out behind the other legs so the palps were unable to extract them. Misaligned legs might have been the horse rather than the cart since the cause of five legs losing their grip is uncertain.

At 20hrs. the animal seemed active and walked well using the legs, only the whips did not function correctly. At that time it autotomized one whip and the exoskeleton sclerotization process was progressing. As seen here, at 30hrs it gave up and died. Few other arthopods would die from such a molt but amblypygids die from far less problematic molts. This first developmental mismolt in nearly three years of keeping this species may have related to random misfortune, the large size of the animal at such an early stage, or because I had written the day before I had never seen an accidental mismolt that could not be pinned on easily identified habitat shortcomings.

While I have seen entomophagus mold and grain mites damage and kill other invertebrates, I have not personally lost a single amblypygid to either (because they are not terribly prone to these issues rather than due to my abilities).

Whipspiders are prone to postmolt death, unlike any other invertebrate kept in captivity. Most arthropods can have molts that result in severe deformities and yet only die because they run out of food or moisture since the deformity prevents ingestion. These other arthropods can experience horrendous, crippling molts and continue on for months. Whipspiders that have bad molts resulting in somewhat minor deformities (such as slightly twisted palps or legs) usually die before the exoskeleton fully darkens in color. Large immatures and adults nearly always die, but very young specimens in the first few instars are less likely to die. If a specimen has any appendages stuck in the exuvium and it survives long enough to darken fully, only at this time should pieces of the old exoskeleton be cut off.

As specimens grow older they become prone to bad molts, seemingly independent of external conditions. This takes place only after postultimate molts no longer result in increased size—two to three years, and two or more molts following initial sexual maturity. It seems to be a function of old age because previous same-sized

A. Immature male *P. barbadensis* wild-caught with obviously regenerated, undersized whip.

B. After three months of feeding it grew quite fat but then autotomized the larger, unregenerated, left whip 24hrs following a botched escape attempt.

C. Thirty days after minor accidental damage resulting in self-autotomization, the male molted to initial maturity and regenerated the fully functional left whip as shown. Other than an exuvium and a huge increase in prosoma size, the previous evening's molt is evidenced by the blue rather than brown ambulatory femoral bands, and unusually light colored maculations on the prosoma. The transformation to near normal color is incredibly fast compared to *Damon* or *Heterophrynus* and likely relates to exposed daytime shelters. Notice the size ratio of the antenniform femora changed. It is likely the extremely rare wild specimen that appears to have regenerated both whips in nature only survived because it lost them at different molt intervals.

molts in the same terrarium, under unchanged conditions, have been perfectly successful.

Immatures and young adults have trouble molting if they are greatly dehydrated or the molting surfaces provided are inadequate. Dehydrated animals can usually split the prosoma but will not get much further and will die even if a few legs break free. Molting surfaces are important but the surface texture is more important than the structure. Unlike mantids and phasmids for which perches in the enclosure must be carefully considered so specimens do not molt too low and damage themselves against the bottom, amblypygids seem much smarter. A percent (based to a degree on enclosure design) of mantids and phasmids choose a bad spot to molt even when there are perfectly adequate perches a few inches away. I have never seen an amblypygid molt too low and damage itself in a similar manner and my sample size is at least a thousand molts. Though whip-spiders do not seem to ever choose too low a surface, they still can fall or lose purchase from inadequately rough surfaces, mismolt terribly, and die. Once at a reptile show I saw a rather distressed, but strong *D. diadema* missing both whips that was brought in by a hobbyist. I asked what happened and the keeper said the whips were lost during the molt a few weeks earlier. A few answers later I discovered it had been kept on substrate with no molting surface, as though it were a scorpion, and apparently autotomized the whips in order to free itself. This animal was a rarity, as such a molt would normally end in a quick death.

This female lost the smaller whip down to the trochanter when she was a tiny 2nd instar and after half a dozen molts the whip is still undersized (presently 36mm body length, 2nd stage adult female).

Various amblypygid species have the ability to regenerate lost or damaged limbs. The antenniform legs known as "whips" grow back readily at the next molt. This is an important ability to possess since whips are important but commonly lost to fighting, bad molts, environmental hazards, and minor damage. Animals can survive in nature with one whip (for evidence see wild *Paraphrynus laevifrons* and *Acanthophrynus coronatus* natural photographs and the following explanations) but animals with no whips are likely doomed to be consumed or starved outside captivity. Wild-caught specimens are also regularly missing whips but these could be blamed entirely on capture and storage methods. However, of the limited number of wild specimens I have seen, there have been a large percentage of specimens that most certainly were equipped with a regenerated antenniform leg, including a number of *P. marginemaculatus*, both recently acquired male *P. barbadensis*, a *D. diadema*, and a *P. mexicanus*.

The regenerated whip is large enough to be useful after a single molt though it is rarely as large as the original. I have personally observed regeneration in dozens of specimens from the listed genera and the whip has always been functional after the first molt, unlike some arthropods that regenerate only a nonfunctional stub at the next molt. An especially small whip will regain some size after a second molt but it does not continue to grow in later molts. The size after regeneration is relative to how far back it was lost and the time until the next molt. The regeneration point is either the joint between the patella and tibia or the joint of the femur and trochanter. If broken beyond either point it will be discarded back to the closest point before regeneration. (Autotomy is controlled by nerve bundles and does not occur on anesthetized animals.) If the terminal tarsal segments of a whip are twisted

3rd instar with regenerated whip lost down to the trochanter at 2nd instar, *Heterophrynus batesii*. Only a single molt was required for observed regeneration.

during a handling mishap and no longer fold correctly, the entire tarsus and tibia will be discarded in a day or two. Sometimes, after fighting, whip tarsal segments are bent back behind the tibia or more proximal tibial segments. This should be carefully unhooked by hand because if the animal cannot move it the whip will likewise be autotomized after a few days. Early instar animals are more likely to autotomize the entire whip than large matures and adults, which usually only autotomize back to the patella. If the tip is caught in a lid the entire tibia and tarsus will be found hanging there the next day, not just the last few articulations of the tarsus that were trapped under the lid. Every so often a specimen will lose and regenerate the same whip a few times over its lifetime for no apparent reason. If damaged at or below the trochanter the antenniform leg may never regenerate.

Regenerated whips can have more articles than the original and Weygoldt (2000) suggests after a few molts they could become much longer than the original. However, he may have based his theory on some function of aberrant wild specimens as whips do not continue to grow larger in reality. I have had a few *Damon diadema* lose whips at 2nd instar and the new whips remained visibly smaller than the original through multiple postultimate molts. (Look closely at the female with 1st and 2nd instar progeny who lost a whip herself at 2nd instar.) The *Heterophrynus batesii* female pictured with young molting off her abdomen lost a whip at 2nd instar and the regenerated whip is still only about 75% of the other even after going through all the molts to attain maturity as well as a postultimate molt. In addition, I have seen a number of *P. marginemaculatus,* one *P. mexicanus,* and two *P. barbadensis* wild specimens that have one undersized (never one that is abnormally large) whip—almost certainly regenerated in nature. Lastly, although the regenerated whips will appear very obviously undersized in photos, it is easy to miss on living specimens because they are held up in the air and are often moving. (I did not realize my breeding *Heterophrynus batesii* female had been the one to lose its left whip at 2nd instar during initial shipping until this photo. In most photographs the difference is not apparent.) Regenerated whips may not always have more articles. A count of tibial articles on a single known regenerated 3rd instar *H. batesii* yielded four less tibial articles, while conversely a rather tiny, certainly regenerated whip from a wild male *P. barbadensis* had twenty more tibial articles than the original (50 vs. 30). Counting articles may be academic since regenerated whips never fully obtain the size of the original independent of article number.

The walking legs do not grow back easily. One immature *D. diadema* lost half a leg on the molt to 2nd instar and molt after molt there was no sign of the leg growing back. It seemed walking legs might not be capable of regeneration despite an article that said walking legs could regenerate, but did not claim or reference a specific example (Ditzel 2002). However, on the molt to 6th instar the leg finally grew back, only a hair smaller than its counterpart. It took me some time checking every specimen and looking closely at this specimen's exuvium and regenerated leg to make sure I was not looking at the wrong animal.

Palps have limited potential for regeneration. The specimen pictured lost part of its palp in a fight. It fed acceptably with one palp but prey had to be smaller than usual for the same size specimen. After the first molt a tiny stump about 5% of full size grew back where the tibia had been, but the associated whip was lost down to the base of the femur and the exoskeleton had to be manually removed. In the subsequent molt the palp and tibia would have grown back to about 30% and still been nonfunctional but the regenerating palp made it impossible to break free of the exoskeleton and the animal died only partly extracted from its molt. In retrospect, if I had originally cut the palp off down to the trochanter it may have prevented the animal from getting stuck in the molt. It is also possible such a procedure might have killed it, since palps may not naturally autotomize, resulting in it bleeding to death.

Regeneration requires resources and likely delays the next molt somewhat. The specimen missing a palp eventually was a year or two behind its siblings in development (and never reached maximum size and seemed stunted), though this probably was the result of a reduced ability to catch prey rather than energy spent in attempting regeneration. I have followed a few broods of *D. diadema* and both of *H. batesii* where one or two specimens were missing one whip at second instar. In all cases the specimen or specimens missing a whip were among the middle to late group to molt but never the last of their siblings. I have seen a few wild-caught animals with a missing walking leg and any delay is hard to demonstrate because development seemed similar to other specimens but no specific molting period reference was available.

Freshly molted *Damon diadema* 3rd instar.

1st instar appendages are growing pink, three days after hatching.

6
REPRODUCTION

Sexual dimorphism is a common feature of the Amblypygi though its exact expression is variable. Males and females of most large species tend to look very different when they are mature due to the male's advanced weaponry or relative body size and shape. In species with long palps, the male's are commonly much longer (when folded) than the femur of the first walking leg, while the female's palps are shorter and stockier. In these cases, the male's elongate pedipalps are thinner and may be curved in the middle or near the "elbow." In some species such as *Phrynus barbadensis,* the palps are similar in length, but the female's may be thicker in cross-section. In smaller species like *P. marginemaculatus* with similar size pedipalps, the mature male tends to have slightly longer pedipalps in relation to the body. Females look similar and are often larger, but the difference is the ratio, not the overall size. Also, in most species the male's abdomen is commonly thinner and narrower in relation to the prosoma, but even this is not a given. With experience it is often possible to learn to identify males visually from above even for species with very subtle sexual dimorphism. This requires much familiarity and definite confirmation of the original groups.

Males possess exaggerated palps primarily for the purpose of sizing up and fighting each other. The exaggerated appendages are used in sparring over territory and

From above it can be difficult to tell which of these females is carrying a broodsac. *Paraphrynus mexicanus.*

females, though there is a slight chance a small male could wander off with a female while two large, old males are busy fighting. Chances of a substandard male fertilizing a female's young in place of more powerful males is highly improbable because courtship periods tend to be longer than battles and the victorious male would

Sexual dimorphism is blatantly obvious in these second stage adult *H. batesii*. Staging this photo required fifteen minutes and a few short chases across the floor. It is difficult to get two skittish, standing adult animals of this species on the same plane, close to each other, and in focus.

Female operculum, *Paraphrynus mexicanus*.

Male genital operculum, *Paraphrynus mexicanus*.

just remate her later anyway, after competing suitors are scared off. Large males do not usually attack small males in captivity and seem unable to recognize their undersized counterparts as competition. However, spermatophores are not observed, so immature males are not mistaken for a receptive female. Older, full-size males may cohabitate in a large cage with multiple decorations, but generally it is safest if only one male is kept with a group of females to prevent aggression. Close quarters also may eliminate or reduce such aggression. Males of species without greatly enlarged pedipalps are almost never seen sparring and are no more aggressive to each other than to a female.

These third stage *Damon diadema* adults are very easy to tell apart.

Adult pair of *Phrynus marginemaculatus*.

First stage sexually mature adults can look rather similar at a glance, *Damon diadema*.

Even in the large species it can be impossible to differentiate male and female during the early instars. It may still be difficult when the adults first reach maturity because they are small and sexual dimorphism has not yet developed to a great degree. For the first two molts following sexual maturity, whipspiders continue to grow. (Later postultimate molts result in no appreciable size increase.) Males not only increase in size but also increase in exaggeration of the gender features, so older males often look incredibly different from females when younger males did not. This uneven type of growth, referred to as allometry, is also seen in the front legs of male *Macrobrachium* shrimp, tail appendages of male *Porcellio laevis* isopods, pedipalps of male *Meglopsalis* sp. opiliones, horns or jaws of various male beetles, and many other arthropods. Of these only the

The exposed genitalia of a male *H. alces* dead specimen.

In some species the male's opisthosoma is small, thin, and squared off, *H. alces*.

crustaceans share the lack of a terminal adult molt that results in multiple variations of gender traits expressed in a single individual over its lifespan. The few whip-spider species in which the male's chelicerae are enlarged or the prosoma is raised behind the chelicerae do not display these features before maximum size is reached. Younger animals usually will display differences in other dimorphic body parts.

Pair (male on left) *P. marginemaculatus*. The genital operculum of the male and female is visibly different but not incredibly so. Much of the difficulty encountered in sexing this species is related to the preponderance of females. These two are not dead specimens, they came in nearly frozen from shipping and fully recovered later.

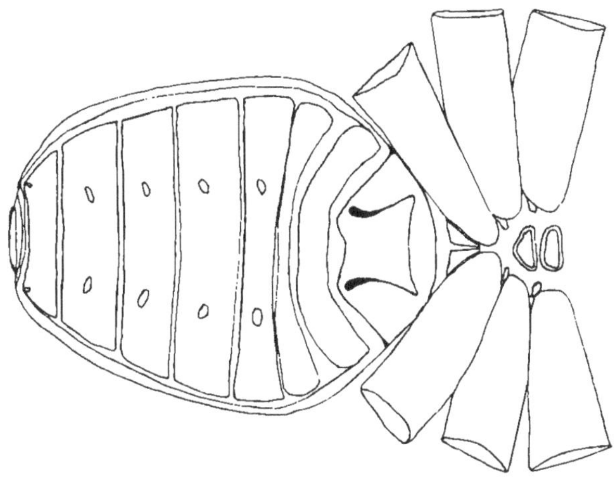

Female operculum diagram, *Damon.*

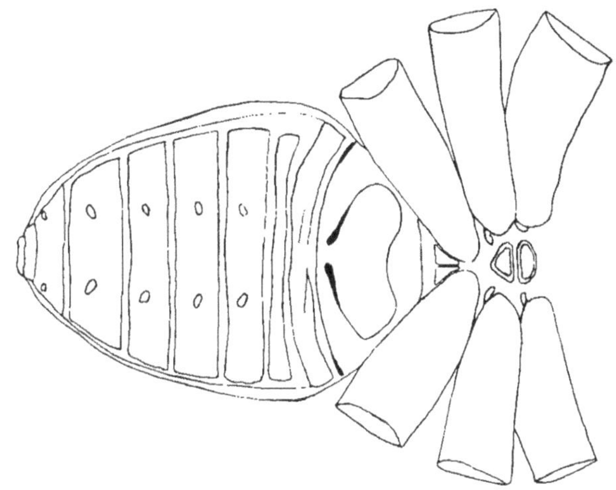

Male operculum diagram, *Damon.*

As mentioned, adults of small species, immatures, and even young adults of large species are often difficult to sex accurately by comparing the relative palp size and abdominal shape. In a number of species the males have notably longer whips relative to body size even before maturity is reached. Sexing with the whips can likewise be difficult since specimens may be of different age or have regenerated whips and moving whips can be difficult to compare. In some species the adult female's ovaries can be seen through the abdominal sternites but it varies by individual and can be confused with the male's internal reproductive organs. Fortunately there is a more reliable way to differentiate the genders. The pedicel which butts up against the prosoma has a very tiny, split sternite that is barely noticeable. Following it, the large, expanded second sternite that covers the genitals on the third sternite is called the genital operculum. It contains the gonopods and the first pair of book lungs. This hardened plate on the underside of the abdomen looks different according to gender, species, and development. For *D. diadema,* the center of the female's operculum has a small trapezoid shaped indentation while the male has a large, heart-shaped marking. *Damon* can be sexed definitively as early as 3rd instar but other genera can still be a little difficult at maturity. The exact shape can vary widely for various genera and species, but the shape and size are sexually dimorphic, with the male's genital operculum usually being larger and more rounded.

Weygoldt (2000) mentions a few sexing methods I have not had luck with but which may prove useful to some keepers. He says there is always a fringe of reddish setae at the apex or open edge of the female's operculum. However, this feature is difficult or impossible to see in the close-up photographs of female genital operculums included in this text, and is even more difficult to see on a living specimen. It can be located on mature *Damon* but the genital operculum shape is easier to see and can be seen half a dozen molts earlier. As an alternative, he suggests anesthetizing specimens and peeling back the genital operculum to look at the gonopods. This

Male genital operculum *Damon diadema*.

Small, first stage adult *H. batesii* female.

Female genital operculum *Damon diadema*.

Small, first stage adult *H. batesii* male.

method should be functional, but for me the risk of killing the specimen or damaging genitalia when it is nearly impossible to acquire most species is unacceptable. Furthermore, the genital operculum differentiates at an earlier instar than the genitalia for at least *D. diadema*. Male genitalia is highly variable and can appear as different between members of the same species as members of different species (Weygoldt 1999). Different genera have different looking reproductive organs and so do different developmental stages, so it may take considerable time and a large numbers of specimens to become familiar with this anatomy. Some species may not be sexable unless you have both a mature male and female to compare because of differences in the female genitalia across the order.

Pairing and Courtship

Once the sexes are determined and the animals are mature and conditioned for breeding through adequate feeding, the next step is the introduction. Mating should not be

Female genital operculum, *Phrynus barbadensis.*

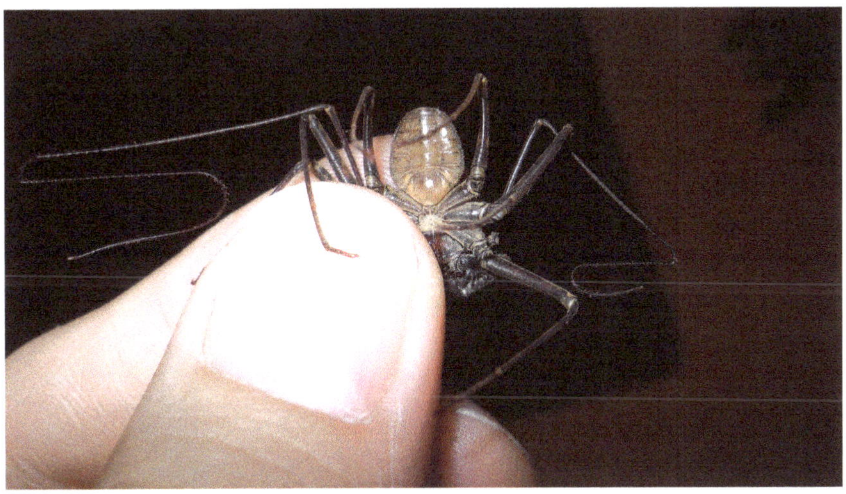

Even this immature male is easily identified by the operculum shape, *P. barbadensis.*

It can be difficult to make out the genital operculum shape from an angle. I have used a small dental mirror with pinned specimens which may be useful with species that are not overly skittish and do not closely hug the surface.

attempted in anything smaller than a 5-gallon (19 L) terrarium (or 2.5-gallon (9 L) for *Paraphrynus* and *Phrynus* spp.). While mating even the large *D. diadema* has been accomplished in a tall 32 oz (1 L) deli-cup, the adults can attack, damage, and even mortally wound each other in such small quarters. Even in large enclosures it is important to offer a number of hiding areas and the initial behavior should be carefully noted. Introductions can take place during the day even for species that otherwise refuse to eat or move at this time. Amblypygids respond to a new animal immediately, so they should be placed next to each other in a large cage to be certain the initial reactions will be observed. Nervous, almost terrified, whip movements are normal in the first few minutes. If the adults are seen boxing or poking rapidly with their pedipalps, they are fighting, not mating. Actual courtship involves touching with the 'whips' and slow, deliberate pedipalp movements. Rapid whip movements can occur first in either case. Fast lunging or rearing movements of the palps mean they do not like each other and probably should not be left together for any length of time. Any extension or use of the chelicerae is unusual and means they desire the immediate death of one another. Aggression may be eliminated by adding more hiding areas, but it is best to move them to separate cages and try a few weeks or months later. If accustomed to each other, a pair or trio can safely cohabitate for a year or more and (with exception of *Phrynus barbadensis* and possibly other species) adults will not harm each other during molts. Cohabitating pairs share a hide most days and are often oriented facing each other at rest.

Pair reintroduction three months after young dispersed, *Heterophrynus batesii.*

Courtship behavior with palps in a V-shape, *Heterophrynus batesii.*

Sexually mature specimens of vastly different sizes can be successfully paired, but size should be matched if possible. A large, older female is more likely to attack a small male than vice versa. Years ago when I was not so certain the initial reactions were significant, I discovered a large female killed a small male *D. diadema* overnight. In retrospect, her initial behavior made this foreseeable but she seemed to settle down quickly after introduction. (Initially she opened up the palps a few times and

Spermatophore formed by a very young male *Heterophrynus batesii*.

seemed agitated.) For observation, specimens may be introduced during the day, but spermatophore production generally only occurs at night.

Behavior during introduction likely differs somewhat across genera and species. A pair of *Heterophrynus batesii* were separated for eight months for gestation and recovery, and then reintroduced five weeks after the female molted. The following behavior was noted:

The pair do not get close but flick their antenniform legs in very rapid back and forth movements that seem like intense nervousness or excitement, within five minutes this behavior stops. The initial, nervous whip movements may have more to do with uncertainty of the new animal than ritualized mating behavior. The male and female take turns approaching and backing away for several minutes. They always face each other. For the next half hour the female raises and lowers her body every few seconds at intervals in the courtship when the two are in close proximity. Her body is close to the substrate and is raised about 1.2″ (30 mm) each time. (Schleser and Roberts 1997, however, report in their observations the male performs "deep knee bends.") Next, the couple holds up the palps in a V shape as they face each other and use their whips to gauge each other. After a few hours of what has become very slow and gentle whip stroking, the male lowers his body to the substrate (usually the surface of wood or cork bark in captivity) and forms a spermatophore while still caressing her with his whips. Whether she comes forward to accept it or not, this same procedure is often followed for the next few nights. Afterwards they rest during the day facing each other within a 1/2″ (12 mm) of each other for the next month independent of which hide in the cage is used.

In *Damon diadema*, courtship and familiar pair behavior are similar in some ways. The whips are also of primary use initially but the animals keep their bodies close to the surface at all times. The palps may eventually open but stay on the same plane as the body. After some time the male performs short twists of his closed palps, through a flicking rotation of the trochanter. He sometimes opens the palps and seems to reach towards her but does not hold onto her as she backs away if he gets too close. After a few hours of face-to-face courtship behavior, he will finally turn around and open his genital operculum against the surface to expose gonopods that form the strange white structure which holds the sperm. He turns around again and entices her forward. After mating, pairs are often found in close proximity even when there are multiple, acceptable perches in the cage.

Paraphrynus mexicanus late-stage courtship shares some common themes as described in the following captive observation. The male and female faced each other at slightly over 1″ (>2.5 cm) distance from the front of the prosomas. Both directed the whips towards each other over a period of twenty-five minutes. During this time the male performed double flicks of the antenniform legs every second or two while the female responded only every five to fifteen seconds. His palps were held up in a V-formation while hers were held in the standard closed position. He then turned around and spent eight minutes closely pressed against the cork bark forming the spermatophore. Another female injected her whips and then decided to invade the territory. The courting female opened her palps fully towards the intruder twice who turned around and backed out the other side. A minute later the male turned around and held his palps straight up as he "licked" the body of the spermatophore from base to tip using his chelicerae. This seems to be a last minute check to make sure the formation is adequate. The check lasted barely a minute, then he backed away rapidly while shaking the whips towards the female who came forward with her palps up in a V-formation. She seemed to know exactly where the spermatophore was without touching it with her whips. In a few seconds she touched the end of the spermatophore with her genital operculum but did not open and grab it. She may have been spooked by a meddling photographer. Her fertilization was verified by ova development so she shortly took up this or a subsequent spermatophore.

Keep in mind that the above descriptions are based on observations of specific mating events and individuals of the same species can differ somewhat in behavior.

This spermatophore formed by the male after he molted and greatly increased in size is the same size and shape as those formed when he was much smaller. *Heterophrynus batesii.*

Mating differences between two females of the same species are detailed for *P. barbadensis* in the species list chapter.

Spermatophores

Males of all species and genera, except one, glue an upright spermatophore (resembling a tiny white golf tee or football goal) to the upper surface of a log or bark in the cage. *Paracharon caecus*, the only member of the family Paracharontidae, is an unusual creature that is not known to form spermatophores (Weygoldt 1999). The whipspider spermatophore is relatively large compared to the adult for something that only transmits sperm, but extensive sculpturing reduces the material required. The structure is generally consistent in size

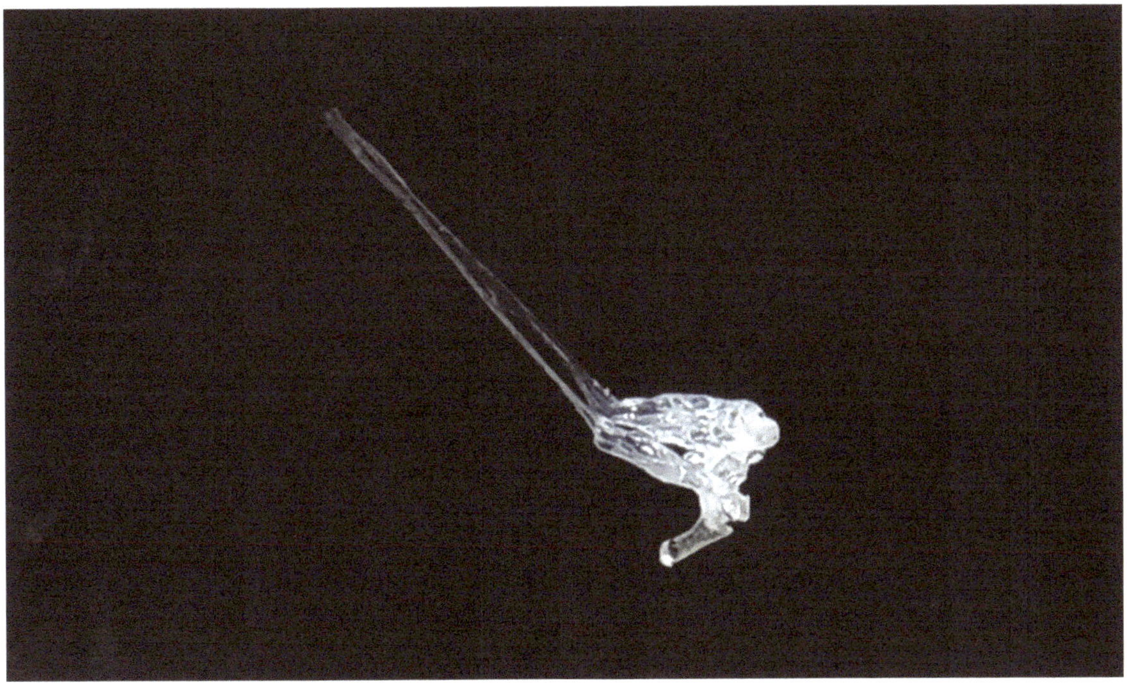

Spermatophore of *H. longicornis* from Pará, Brazil. © César Augusto Chaves Favacho

Early maturity and reproductive capability are somewhat fluid propositions as this *H. batesii* 7 mm spermatophore was found on June 23, 2013, from a 6th instar F2 male that molted a month or two earlier and hatched out the prior September 31 (as little as seven months).

and shape for each species, independent of the size of the individual male. The spermatophore is constructed on a vertical or angled solid surface and, in the families discussed here, cannot be formed on a substrate made of small particles. It will be oriented perpendicular to the surface, or at a specific angle, so that it can be perpendicular from the front and at a forty-five degree angle from the side. On close inspection the structure is amazingly complex as even the simple basal stem is angular in cross section rather than round. While it displays bilateral symmetry, the components look different from multiple angles. The specific size, shape, and angle are designed to fit the female's genital organs so getting her to take up a spermatophore from a different species may be impossible even if they were switched between cages and somehow artificially enticed to take up the spermatophore.

The length of the spermatophore is often relative to how far the female can stand off the substrate. *Damon diadema* and *Heterophrynus batesii* are not terribly dissimilar in body mass but *D. diadema* have spermatophores only around 0.24" (6 mm) high, while *H. batesii* spermatophores are 0.83" (21 mm) tall. The much smaller *P. marginemaculatus* spermatophore is relatively large at 0.16" (4 mm), but is formed at a lower angle to accommodate the female, while that of *P. barbadensis* is 0.26" (6.5 mm) and formed at an even lower angle. I had imagined the spermatophore of *Paraphrynus mexicanus* would be similar to *Heterophrynus* in form and the male does construct a relatively thin, long 0.31" (8 mm) spermatophore. However, the single observed structure was at an amazingly low angle. While the former listed species

Spermatophore stalk is temporarily left behind after the female tears off the end with her operculum, *Heterophrynus batesii*.

Spermatophore compared to adult, *Phrynus marginemaculatus*.

Spermatophore of *Phrynus marginemaculatus*.

Early courtship or aggression? *Paraphrynus mexicanus.*

angles were based on multiple first-hand observations of spermatophore structures, this is the only *P. mexicanus* spermatophore I have observed.

The male manipulates the female into taking up the spermatophore through long periods of delicate courtship primarily consisting of antenniform leg movements. Males of many species do not grasp the female's pedipalps or aggressively position her over the spermatophore. The female moves over the spermatophore, opens her genital operculum and moves down until the end of the spermatophore is inside. She then closes her operculum and tears off all but the base. The male is very aware of a receptive female and does not make spermatophores if there is not a sexually mature female nearby that shows interest. Males kept by themselves for years never make a single spermatophore (and unmated females of the species discussed here likewise live for years without ever forming a broodsac). However, the same males will make spermatophores within hours of encountering a female capable of breeding. If the female is mature but not at a point in her developmental cycles where she is capable of making eggs, he will not form a spermatophore. Although he can judge a female's suitability with perfect accuracy he cannot determine how many spermatophores she can use. She is likely only capable of taking up one spermatophore, but the pair will continue courtship gestures for one to three weeks. In the presence of a

Note the very low angle of this *Phrynus barbadensis* spermatophore, female 2, March 1, 2013.

The base of a *P. barbadensis* spermatophore can be seen in the shadow of the first walking leg. The main portion was torn off by the female's genitalia, female 1, March 25, 2013.

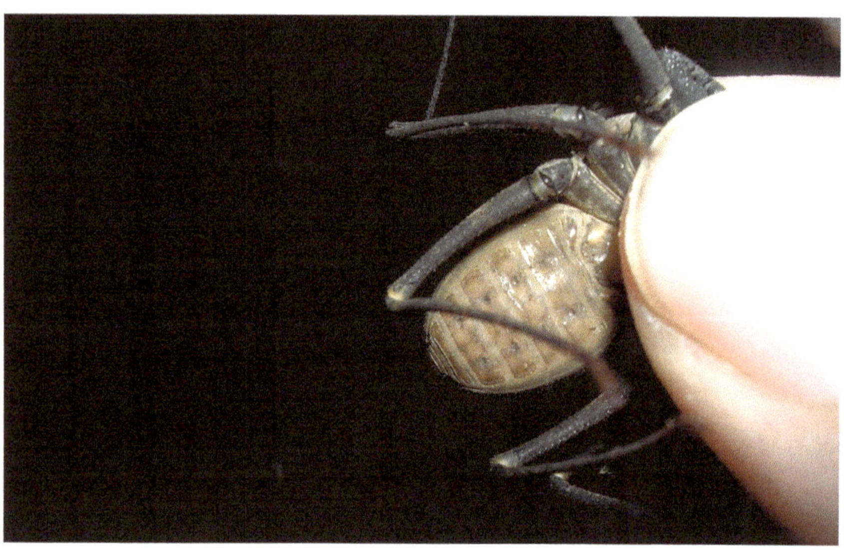

Eggs became visible through the ventral sternites of female 2, May 27, 2013. Brood-sac extrusion took place on June 11.

suitable and receptive female he will construct three to ten spermatophores over this time.

The presence of signs of spermatophore production does not always result in a reproductive female. A successfully mated female will almost always display a notable increase in appetite within days. If a spermatophore is found intact the following day

Spermatophore, *Damon diadema*, this particular structure is more thickly built than usual but is the approximate dimensions and shape of standard.

it is usually just an extra, though it could signify she has refused the spermatophore. Very rarely, on some surfaces, the entire spermatophore will rip off and the base of the stalk may be seen sticking out of her operculum. (I observed this when I used plastic yarn crafting material, since the surface was solid enough for formation but the spermatophores did not properly adhere.) It is possible to confirm mating when the base of a broken spermatophore is found still attached to an object but this is not common. However, unused spermatophores and broken stalks are normally eaten down to the bark the following night if found during the day, or within an hour if observed during daytime courtship. Usually the male, but sometimes the female,

Eggs can be visible through the intersegmental membrane as large, green spheres. *D. diadema.*

eats it. It then becomes impossible to tell where it once was. Her spermathecae, according to captive results, do not hold sperm longer than five months. If a fertilized female molts or does not have eggs within three to five months of mating (usually less than three), she will need to be mated again.

Parthenogenesis is reported for a few small *Charinus* species but has no bearing on other species being kept in captivity. In such cases there are no males and therefore no spermatophores. *Charinus acosta* from Cuba is a miniscule creature, about 1/16" (1.6-2.3 mm) across the carapace, which averages four offspring per broodsac and is the only whipspider in which males are unknown (Armas 2005). Populations of the similar Old World *Charinus ioanniticus* are found in Egypt and Turkey as male and female, but in Greece only females are known (Weygoldt 2000).

The Broodsac

Within a week or two of mating the developing eggs often become visible through the intersegmental membranes on the underside of the female's abdomen or beneath

This broodsac is being extruded, *Damon diadema*.
The eggs are not out of focus, this is the way they look.

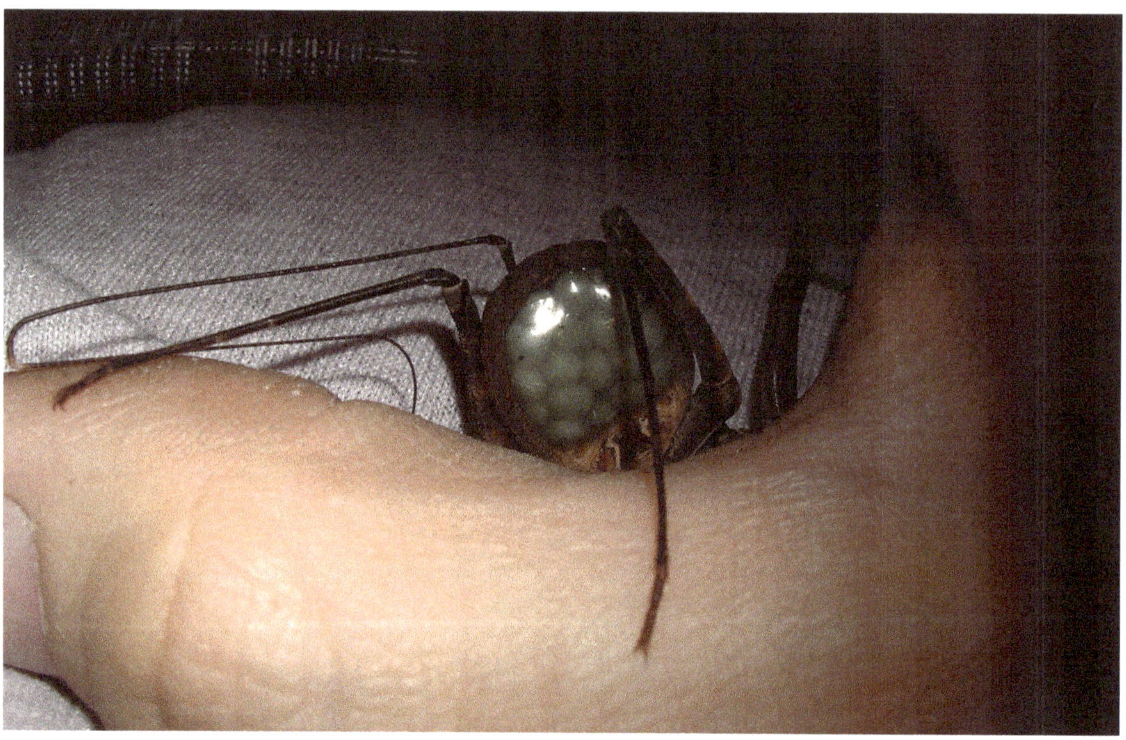

The freshly extruded broodsac flattens out in the bowl shaped abdomen as long as it remains inverted, *Damon diadema*.

First female with fresh broodsac, notice one leg already dropped off, *Heterophrynus batesii*. The eggs are not out of focus, they appear fuzzy in real life.

The broodsac of a healthy female transforms to a chocolate brown after four days or so, *Heterophrynus batesii*.

Postembryos are evidenced by the change of shape, *Heterophrynus batesii.*

1st instars are pale 24 hours after hatching, *Heterophrynus batesii.*

The broodsac opens to expose postembryos for hatching.

the sternites if they are see through. This can be difficult to make out and is *not* the extruded sac. Ova will be pea green and large; tiny, white, oval objects can be seen on the ovaries of unmated females and are not developing. At this point the female is fertilized but the eggs are not, which is why the male usually sticks around for weeks or months in nature to ensure paternity. The eggs will not be fertilized by sperm stored in her spermathecae until extruded externally—another male could replace his sperm during this initial period of development. The growing eggs are often easy to make out in *Damon*, *Paraphrynus*, and *Phrynus*, but *Heterophrynus* sternites and intersegmental membranes are fully opaque. Still, even in *Heterophrynus* the development is seen as a rapidly swelling opisthosoma.

The ova may be externally visible within a week or two of mating but it will take five to eight weeks for them to become too large to remain in the opisthosoma. This can take a few months or more if the female was not well-conditioned. At this time they are extruded into an external sac that is still connected to the female's genitalia. A few days before this happens the small tail segments are intermittently extended. Less than 48 hours beforehand, the ventral outer edges of the female's opisthosoma extend into a narrow lip that will hold the sac—these are the pleural folds. The exact shape and deformation varies as the *Heterophrynus* abdomen barely holds the eggs, while some of the savanna *Damon* like *D. variegatus* extend exaggerated pleural folds which cover most of the broodsac membrane as a second layer of protection. However, in all cases the abdomen deforms notably in preparation for extrusion, so any folding expansion of the pleural membrane makes it easy to determine when extrusion of the broodsac is imminent.

The broodsac is extruded (the eggs never touch the air and the base remains

Phrynus barbadensis, freshly extruded broodsac.

Paraphrynus mexicanus. First broodsac of five females, 15-25 eggs each (June 13, 2013).

Postembryos ready to hatch into 1st instars.

Is this fat animal a female with eggs inside or with pleural folds over the edges of a broosac? In some species it can be hard to tell if the eggs are internal or if the broodsac is obscured unless the animal is flipped over, *P. exopthalmus*.

Sclerotized broodsac, *Phrynus marginemaculatus*.

Postembryos are evidenced by the nonspherical shape even though the membrane is opaque, *Phrynus marginemaculatus*.

firmly attached below the genital operculum) into a round glob that flattens out into the edges of the pleural folds while the female holds her opisthosoma in a horizontal and upside-down position (*Damon* and *Phrynus* spp). The female *H. batesii* hangs with the prosoma upwards (abdomen perpendicular to the ground) during extrusion and about six hours later switches to the upside-down position—it may stay that way most of the day for weeks. The small tail segments of the opisthosoma that serve an important role in molting are nearly always extended during broodsac extrusion but do not play an obvious role. Ova are essentially extruded at the same time but the process takes about two hours before the broodsac is the approximate final size and shape. If a female is disturbed during

Freshly extruded broodsac, *Phrynus marginemaculatus.*

extrusion the shape can become deformed when she changes position and hangs the wrong way. This is a very terrible time to disturb the female as she will lower her body towards the substrate and move around when she should not. The surface starts out very soft and sticky, so substrate materials that contact it can become permanently embedded or make holes. A deformed shape usually does not lead to death of the brood but damage at this time can cause an abortion within days. At first, the broodsac membrane is clear and the green eggs are visible. Excess fluid within the broodsac appear to be absorbed by the female's genital connection over the next three or four days because she rests upside down and the broodsac decreases in mass. (If the eggs absorbed the fluids only the shape should change.) As the fluids are withdrawn the shape of individual eggs will become notable on the broodsac surface. Over the course of a few more days the membrane hardens, dries, and later turns golden brown to black. Once the membrane fully darkens it is usually still possible to make out the individual eggs, but the total number cannot be counted since only the outer layer is visible.

In the Phrynichidae and some Phrynidae like *Phrynus parvulus*, it is not possible to see the eggs from the side as the pleural membrane folds down to cover the sides of the broodsac. In some Phrynichidae found in dry areas such as *Damon variegatus,* the female's abdomen expands so much that her pleural membrane entirely covers the outer eggs and all but the center of the broodsac. (In *D. diadema* and most *Phrynus* the folds are small and only cover the leading edge of the outer eggs.) Large pleural folds that cover much of the broodsac membrane are almost certainly an adaptation designed to shield the developing embryos from desiccation, but would also protect the thin membrane from physical

1st instar *Phrynus marginemaculatus* on mother's opisthosoma.

Sternites are see-through in some species and the ovaries are sometimes visible. These tiny white, infertile spheres look quite different from the large green developing eggs. *P. mexicanus*, Nov. 2012.

damage. The broodsac is still very noticeable with a look at the underside of the opisthosoma, but the females do not flash the underside of her abdomen, especially in these species. A small desert species of Phrynidae, *Phrynus asperatipes,* from the Sonoran desert in southern Baja California, Mexico, has pleural folds that are even more extensive. The folds meet at the center and the only evidence of the broodsac is a line down the middle (and lack of any visible sternites). In most Phrynidae the abdomen may curl slightly around the broodsac but it is still visible from the side, sometimes even sticking out from above. It can be less visible on some *Phrynus* but it is impossible to miss on *Heterophrynus*. In any case, the female still eats and behaves normally when holding eggs. When eggs are not easily seen from above, females holding broodsacs often go unnoticed for months by beginning keepers who pick up wild-caught animals. Most or all reports of short broodsac gestation are probably the result of the keeper not noticing the broodsac. Of course the cause of reporting errors cannot always be known—one old report by Levi and Levi (1968) states the female of our Florida "*Tarantula marginemaculata* . . . carries eggs and young for four to six days" rather than three months. This certainly only refers to first instars but even that brief period is nearly twice as long. This unfortunate error (and the horrible common name) has been copied forward into some more recent guides such as Evans (2007) who says, "The eggs hatch in about a week."

Note ova development within abdomen less than two weeks after fertilization. *P. mexicanus*, April 2013.

Male *Paraphrynus mexicanus* using his chelicerae to conduct inspection of his recently formed spermatophore.

Female *Paraphrynus mexicanus* lowering her operculum (still closed) onto a spermatophore.

While not necessary, separation of the gravid female seems to be helpful and is suggested. Her food intake, moisture, and health can be monitored. A day or two before hatching, the broodsac swells notably and the female holds her body further from the surface; this is the time to remove her if she has not already been separated.

Temperature may be the most important factor over the long gestation period. From the day of mating to the day of hatching the total time required is approximately six months for *Damon* and *Phrynus* and eight months for *Heterophrynus*. Schleser and Roberts (1997) reported an unbelievably fast development time for *Heterophrynus* that seems in error but may result from the high reported gestation temperature of 80° F (27° C). However, I have tried to speed up *Damon* gestation in the past with high temperatures and the few that did not fail horribly still did not develop at a notably improved rate. Furthermore, my first *Heterophrynus* attempt that ended in dismal failure was at 80° F (27° C) and the successful attempt was closer to 72° F (22° C). I purposefully dropped the temperature in reaction to the failed attempt. I had experienced severe gestational difficulties caused by high temperatures in other species, but the first time around I followed the advice of Schleser and Roberts (1997) and associated Amazonian with hot. Daytime temperature in a lowland rainforest may be around 80° F (27° C), but amblypygids stay hidden during that time and may only experience temps as high as 75° F (24° C). If *Damon* or *Phrynus* are kept above 75° F (24° C) it can upset the development of the eggs. At high temperatures a few of the eggs may turn black; this does not necessarily kill the other eggs, but is often a sign of impending doom. Moving the female to a cooler area may help, but usually a single black egg means the entire brood will fail even though the female still holds the broodsac for four months. Keep in mind that temperature is not an absolute factor unaffected by other variables, so different aspects of the captive habitat may affect allowable, successful gestational temperatures through differences in airflow, humidity, and possibly food intake or other less obvious reasons.

The broodsac has darkened to a golden brown from days four to six. *Phrynus barbadensis.*

After a few months of excessive temperatures and high humidity the eggs could grow moldy but they do not because the usual cause of mold is low humidity. Low humidity as the cause of mold may seem counterintuitive, but healthy eggs never mold in high humidity. Low humidity damages the eggs because the thin membrane allows for evaporation and the dead eggs then become open to fungal infection as soon as moisture is restored. If the humidity is too low and mold is seen, the damage to eggs within the broodsac will be irreversible. The exact humidity percent is relative to airflow, temperature gradients in the habitat, and species of whipspider, but

Early extrusion. *Heterophrynus batesii.*

much below 60% is usually harmful, especially if there is no water dish or misting regimen for the female to replenish moisture. A physically compromised broodsac will be discarded in a week or two and generally will not have time to become moldy first. A broodsac damaged by low humidity will often be discarded after it becomes moldy, sometimes months before the time it would normally hatch. On the other hand, if high temperature is the only problem, the eggs will not mold but be carried to full term and not hatch or hatch unsuccessfully and die. Females forced to incubate at high temperatures tend to become sickly themselves and die within the next six months, while those kept too dry for proper gestation (but not too dry for adult survival) are not damaged by the ordeal and normally survive to produce subsequent, healthy broodsacs the following year. (If the broodsac is lost in the first week or two the subsequent molt and ability to mate again are cut short by a few months.)

Females kept overly dry during gestation will regularly lay flat against the substrate, a behavior rarely seen in animals without broodsacs. This only occurs in dry conditions as even the overly thigmotactic *P. marginemaculatus* are more often found with their ventral surface against the bark than the substrate even when touching on both sides. The behavior seems to be an attempt to protect the broodsac from overly dry conditions by limiting the exposure to air. If this behavior is noticed there may still be time to increase humidity or water availability for successful gestation. There are two alternative reasons animals with or without broodsacs will lay flat on the substrate. One, they have been harassed too many times in one day by the keeper and refuse to return to the perch/cork. Two, the perch provided is unsuitable.

Extrusion thirty minutes later.

To feed or not to feed is the question. Considering the very long gestational period, it is important to keep the female well fed during initial development of the eggs before they are extruded. (Still I am afraid to feed my recently mated female *H. batesii* one more cricket since she will not stop eating and her abdomen looks like a droopy ping-pong ball.) A few weeks after the broodsac is extruded, a female's appetite will return. Her appetite will be reduced but usually is still greater than an older, mature male of the same species. After the broodsac has fully sclerotized (one to three weeks) I have always fed the mother a medium-sized prey item two to four times a month up until the eggs are ready to hatch. In 2008 I learned Weygoldt (2000) suggests females not be fed at this time or the broodsac can fall off (no specific details provided). This might be possible with excessive feeding but I never experienced this with multiple brooding females over more than a dozen years and believe there was another factor in the failures he reported. Still, I no longer feed females holding a broodsac as much as they can possibly consume as I did before reading about his experience. When prey is offered to a female holding a broodsac, it should be removed the next evening if uneaten. Large, aggressive prey could possibly damage the broodsac membrane, though I have never heard of such an event. Also, the female often kills active prey without eating it if she is not hungry.

Female holding broodsac upside down the next day (day two). Note substrate stuck to membrane due to a meddling photographer.

Full extrusion after about two hours total.

By day four the excess fluid has been withdrawn and the broodsac shape reflects the outer eggs.

1st instar appendages developing pink coloration three days after hatching, *H. batesii.*

Heterophrynus batesii molting to 2nd instar. This picture was taken at 9 PM, Sept. 9, 2012, after they had been molting for more than an hour.

The dispersal molts are nearly completed at noon on Sept. 10. A large mass of exuviae attached to the mother is a common sight near the end of the dispersal molt.

If the female dies the eggs are doomed. The broodsac membrane is not detached from the oviduct. It is not possible to gestate eggs without the mother. According to Frye (1992), in the whipscorpions with a similarly formed and connected broodsac, the eggs are nourished with "false scorpion milk" as the broodsac is linked to the mother's blood cavity (haemocoel). Eggs are never exposed to air and gestation is functionally similar to live bearing. Although there is no harm in trying to save a cast off broodsac, no success is likely or has ever been reported. Even if possible, 1st instars would need to be introduced to the mother's back, which may not work. If the mother is alive and the broodsac is cast off due to a small portion being damaged or moldy, the eggs or postembryos cannot be salvaged. Likewise, if the mother dies the eggs cannot be saved.

Hatching and Beyond

After three to five-and-a-half months the sac begins to crack and the postembryo stage can be seen briefly. If the broodsac is closely observed, it is possible to notice the eggs changed over to postembryo halfway to three-quarters through gestation. The perfectly round eggs change shape but it is difficult to see much detail through the opaque shell of the broodsac. The cracking broodsac shell exposes the postembryos briefly, which then hatch into 1st instars that cling to the female's abdomen. The young cling tightly to both the upper and lower surfaces of her opisthosoma as she holds her body off the surface. They do not eat at this instar and have very limited mobility. The female should not be fed and any prey remaining from previous feedings should be removed. In the case of large broods, a few immatures can end up falling to the bottom of the cage when they first hatch. These can be carefully placed on the mother's back—despite seeming very awkward and deformed on substrate, they navigate and hang onto her abdomen with expertise (McMonigle 2005). If left on the cage bottom they will die within a few days.

The 1st instars are white when they first hatch. Over the next ten days (plus or minus) they gradually develop a deep green opisthosoma while the prosoma and appendages turn pink. These darken each day until they finally molt to 2nd instar, so color is a loose indicator of the impending molt. Disturbances at this time do not have much effect, so a female with 1st instars can be held as long as she is not dropped. (Who knows what incredible devastation that would cause . . .) Many other predatory arthropods may eat their young if disturbed. Although the eggs were extruded together, the 1st instars only molt one or two at a time as they drop to the end of her abdomen and shed the exuvium. The *Heterophrynus batesii* brood pictured contained twenty-five young and took just over sixteen hours from the first to last molt. *Phrynus marginemaculatus* usually take just a few hours, while *D. diadema* can take twelve or more hours. Eggs are laid together in the broodsac at almost the same time, so it is curious what mechanism is used to delay the molting immatures. If they all tried to molt at the same time they would have molting troubles or knock each other off. Once all the molts are complete, there is a large pile of exuviae all attached to each other and the end of the female's abdomen. This is knocked off by the female shortly after all the offspring have molted. There is often a limited amount of broodsac residue that remains on her sternites until she molts.

Notice the 2nd instars of this species have remained on the female's back after molting (1st instar exuviae littering substrate). They began to disperse a few days afterwards, *Sarax* sp. ex. Cebu, Philippines © Michael Seiter

Now that they have molted into fully functional 2nd instars, they have full mobility of the appendages which is shortly followed by an appetite. The 2nd instar body structure looks very much like the adult but on close inspection the walking legs are outfitted with tufts of extremely long setae and the palps are shorter relative to the size of the prosoma (even on *Phrynus*). The entire brood should be moved to a habitat immatures can be kept in for at least the first four instars (as long as the enclosure is adequate for rearing, it is safer and easier to remove the mother) and will need a small meal within a few days. There

1st instars are very pale the day after hatching, *Damon diadema.*

Molting to 2nd instar, *Damon diadema.*

Paraphrynus mexicanus early broodsac extrusion, second female, just after noon on June 17.

are exceptions—in at least one tiny species of a genus never kept in the states, *Sarax*, the 2nd instars remain on the mother's back for a few days or more after the molt. In such species the mother should not be removed until they naturally disperse. As mentioned, while it may be perfectly safe to leave the young in with the mother of *D. diadema* and *P. marginemaculatus*, I have reared various broods of these with nearly perfect initial survival rates and have always removed the female the day after the young dispersed. Observation of mother and progeny interactions can be interesting but there is no guarantee every mother of a given species will show the same restraint. (There are many animals, including various tropical fish, that display advanced brood care and yet stressed or unsuitable females will eat their babies from time to time.) Also, at a certain age, following a number of molts, young will no longer be tolerated even if the mother is not hungry. Furthermore, some species might not be tolerant of their young under any circumstances. Unless the primary goal is observation or documentation of interactions between mother and young, it may be best to separate them. Keeping the mother and young together can add a whole layer of difficulty, primarily related to prey. Crickets small enough for the immatures to feed on are a little smaller than the immature whipspiders and still present clear and present danger. On the other hand, crickets large enough for the mother to eat may be a hundred times the mass of a 2nd instar and have jaws strong enough to kill or maim them even when they are not molting. Female prey leftovers may result in rancid conditions or mite outbreaks. If mother and young are to be cohabitated, the type of prey being used should be carefully considered and attention paid to cage hygiene. Lastly, after her long, energy-draining ordeal she deserves some rest and needs to feed and rest in preparation for the standard post-reproduction molt.

Within three to six months of her immatures dispersing, the female *Damon* or *Phrynus* will molt. The female *Heterophrynus* is much faster and at 72°-74° F (22-23° C) can molt within fifty days or so following 2nd instar dispersal. Females will not mate before this time but can be ready to mate again within weeks of the molt. Unmated females do not produce eggs, so it may be best to keep pairs separate for at least a few months following this molt to give females extra recovery time. From one mating to the next following egg production (successful or not) the duration is usually a minimum of nine months. Though earlier introductions and more rapid turnaround are possible, attempts at speeding up the time are likely to result in failure.

Reproduction Recap

1. Gender can usually be determined readily at maturity through comparison of the genital operculum.

2. Aggressive behavior only gets worse at night. Do not keep specimens together if any notable aggression is seen.

3. High humidity is always helpful, but is paramount during gestation and the first few molts.

4. Females should be fed regularly when holding eggs and can be separated from other adults any time before young molt and disperse.

5. Older young cannibalize each other—separate immatures after the fourth molt (5th instar) for maximum survival.

The tarsal claw of the whipscorpion resembles that of scorpions but there is also a large tibial spur or spine. In this mature male *Typopeltis stimpsoni* from Japan, the spur is enlarged and used in copulation.

7
PREY AND FEEDING

WHIPSPIDERS FEEDING in nature are rarely documented and then only when it seems unbelievable, incredibly agile, or gruesome. *Heterophrynus cheiracanthus* has been documented catching freshwater shrimp from streams in Tobago. It places only the whips in the water to locate prey while the palps are held at the ready (Ladle and Velander 2003). Amazonian *Heterophrynus* have been seen catching moths in mid-flight, again using the whips, but held aloft, to determine the prey's location (Weygoldt 2002). *Phrynus parvulus* in Costa Rica has also been documented catching moths from the air (Hebets 2002). In Puerto Rico, *Phrynus longipes*, a large and long-legged member of the genus, has been known to catch and eat lizards from the genus *Anolis* (Beccaloni 2009).

In captivity the basic concern is finding the most functional prey, not the most dramatic. Captive specimens have most often been fed entirely on crickets, less often on certain types of cockroaches, for the entire life cycle. Crickets are the most common in use and can be the easiest to monitor, but they eat molting amblypygids and present some challenge to the keeper. Amblypygids can be reared entirely on the live-bearing *Schultesia lampyridiformis*, firefly cockroaches, although a certain percent of the offered prey will learn to avoid capture and may not be found even if the amblypygids grow very hungry. Egg-laying cockroaches in genera that include *Blatta, Eurycotis, Dorylaea,* and *Periplaneta*, can be used in conjunction or as a singular diet. Fruit flies can be a useful supplement to the diet for the first few instars depending on the species and size, but amblypygids will not develop properly if the diet is only fruit flies. House flies are also limited in usefulness due to nutritive deficiencies, though gut loading may counteract this. Other than through light trapping, moths have little to no practical application in captivity due to

5th instar *Heterophrynus batesii* feeding on a large female *Pyralis farinalis* moth. It was not caught out of the air but stalked after landing.

Hunting posture 1, *Heterophrynus batesii.*

Hunting posture 2, firefly cockroach has now been captured.

cost, difficulty in handling, and time constraints of rearing. Wild-trapped insects are not recommended due to widespread use of residual pesticides, lack of repeatability, conservation issues in some locales, and ease of acquiring cultured prey items.

All the species discussed here will consume active, high profile foods including crickets, a variety of non-burrowing cockroaches, moths, and flies in captivity. They almost never feed on burrowing creatures or those that cling tightly to surfaces. Climbing prey may need to be kept in check (i.e., out of the habitat corners) by applying a thin layer of petroleum jelly to the upper edges of the cage to prevent climbing. Whipspiders are far more likely to die of starvation than feed on food that hides, which includes most feeder roaches, mealworms, isopods, etc. They may probe holes with the whips but do not scare out prey. Unlike mantids, whipscorpions, and others that become excited to hunt by movement and often immediately catch appropriately-sized prey which bumps into them, in whipspiders, contact causes appendage retraction and escape response. Prey that can be overly active and erratic, such as *Pyralis farinalis* meal moths, will not be hunted until it lands and stops moving. Unlike many predators that lose interest the moment prey ceases movement, whipspiders

Adult female *Heterophrynus batesii* feeding on a large, subadult *Dorylaea orini* cockroach.

H. batesii late instar feeding on a firefly cockroach adult. Despite the close relation to natural prey like crickets and cockroaches, phasmids and mantids almost never attract interest.

prefer to go in for the kill after movement has stopped.

I long considered amblypygids overly reclusive and somewhat frail feeders because *P. marginemaculatus* and *D. diadema* would wait a few minutes to many hours before beginning to stalk an introduced prey item. Some will wait until evening. (Turning off the lights during the day is not enough.) They usually will not hunt in the keeper's presence and commonly toss away partially eaten prey if disturbed, sometimes even if looked at too closely. Other *Phrynus*, *Paraphrynus,* and *Phrynichus* have proven similar. If the lid is opened or the cage is bumped, uneaten prey is commonly thrown on the ground. Even if the glass is perfectly clean it can be a challenge to get a decent photograph of these species feeding as movement and focus light can cause the whipspiders to toss their food. However, if a cricket is thrown in with a hungry *Heterophrynus batesii,* this species often stalks and catches prey in less than thirty seconds and it usually holds onto prey despite relatively major disturbances. (Nevertheless, it is probably impossible to hold a specimen directly without it first tossing the prey.)

Whipspiders are nocturnal or crepuscular in nature, though in captivity some will eat any time prey is offered. Other nocturnal arachnids like whipscorpions and windscorpions commonly eat prey during the day if offered at that time in captivity. If prey is placed with *P. mexicanus* and *P. barbadensis*, it will not be stalked or eaten until nighttime, with little exception. *Damon diadema* often stalk food after a little while when hungry, and *H. batesii* do not have to be very hungry. I had wondered if refusal to eat during the day was a behavior unique to certain species, related to habits or relatively smaller eyes, but since I have only seen the first two as wild-caught and my experiences with the second two are almost entirely specimens born in captivity, there is a chance it is learned behavior. I recently observed one wild *P. mexicanus* feeding a half hour before the lights went out, and a recently mated female *P. barbadensis* eating a large *Eurycotis decipiens* nymph offered in the middle of the afternoon.

Whipspiders can be choosy and are able to detect and respond to subtle differences in prey. Freshly molted cockroaches and crickets are recognized and picked off much more rapidly than sclerotized prey. I have offered early instar *Dorylaea orini* and *Eurycotis decipiens* to various whipspiders to supplement available prey because it is difficult to grow or even buy and store the optimum quantity and size of cricket. (Even if no crickets die or cannibalize, they grow rapidly.) These two brightly marked cockroaches have chemical defenses to protect them from vertebrate predators and ants, but are generally edible to most large predatory invertebrates. On the other hand, *Deropeltis paulinoi*, an African cockroach that produces a more powerful defense (a yellow, liquid quinone secretion) is avoided by many arthropod predators after they have eaten or tasted one. This species has been offered to my whipspiders a few times because it is very productive. They eat them very slowly over time as long as the roaches do not outgrow the target prey size, and as long as the amblypygids are not fed much else. I had two dozen of these in with an *H. batesii* group of eighteen 3rd instars for a few weeks, but crickets were offered every few days and eaten preferentially. When the whipspiders encounter the *D. paulinoi* they slightly raise the pedipalps, but when a whip touches this prey they relax and walk off. At this point a dozen similar-sized *Nauphoeta cinerea* nymphs were added to test their ability to differentiate a similar-shaped prey item with similar profile and movements. I expected they would likely ignore the new prey but hoped they would discover over time that these were more acceptable prey (there was still one small cricket left from a few days earlier). Within a minute, four of the whipspiders had caught and were each consuming a nymph. This result was easily as good as what I would have expected with crickets. Since the texture of the *Nauphoeta* exoskeleton is slightly smoother than *Deropletis*, it is not possible to say if recognition was tactile or olfactory. Also, this result should be viewed with the understanding that new prey item types generate the most interest. The enclosure can have accessible crickets and four different types of roaches, but when a fifth species of cockroach is added, a few of the new introductions can be caught within minutes. This is not simply because new introductions have not established hiding spots, since established prey are often exposed, and usually none of the new introductions are caught if "new" prey is the same species as one already present.

There is a commonly recited myth that the praying mantis first bites off the head of insect prey to stop its struggling and pain, but this is not supported by direct observations (McMonigle 2013a). The mantis often catches prey head first, but if it catches it sideways or backwards, feeding begins on whatever part is closest. However, my experience is this behavior is true for amblypygids. I have watched countless whipspiders manipulate prey caught backwards or sideways so that the head will be removed first as the first bites are in or near the soft integument of the neck. This behavior is relative to insect prey—I am not certain this would be the same for a vertebrate with no antennae or certain arthropods with an indefinite head region, since other amblypygids are not attacked this way. I have not tested vertebrate feeding responses since they can cry or scream, whereas invertebrate prey seems unaware even before the head goes missing.

Size is important as tiny prey is ignored and overly large prey initiates a flight response. The body of a chosen prey item should be approximately the length of the prosoma of the amblypygid it is being fed to. Whips gently tap the prey from each side to measure it before and sometimes during capture. Prey items as small as the chelicerae are ignored. Oversized prey may be captured by hungry individuals but the excess is commonly discarded. Various isopods

This wild *Acanthophrynus* is feeding on a kadydid in Michoacan, Mexico. © Eddy Hijmensen

Fat *Phrynus barbadensis* immature.

are great as a "cleanup crew" because there are always leftovers, though refuse is usually just a small ball consisting almost entirely of chewed up exoskeleton. Still, I have seen many half-eaten crickets. Large species may occasionally eat small lizards in nature, but it is difficult to get them to accept vertebrate prey in captivity due to size. Dixey (2002) reports feeding tiny anoles and frogs to *D. diadema*, but inability to get them to show any interest in the far bulkier pinkie mice. Immatures of various species will accept fruit flies and hatchling or early instar crickets. Adults feed well on 1/4″ (6 mm) to 3/4″ (19 mm) crickets, as well as large nymphs of egg-laying cockroaches such as those in the genera *Eurycotis* and *Blatta*. Most commonly available burrowing roaches are gigantic as adults, must be offered as small nymphs, and are rarely used with success due more to behavior than size.

The prey (*E. decipiens*) has had its head eaten first.

Scavenging behavior is reported but is unlikely to be useful in proper husbandry. Weygoldt (2000) reports some *Charinus* will feed on rolled oats. I have tried over and over to offer dead, dying, or incapacitated prey of various sizes and species to *Heterophrynus* and other genera with absolutely zero response. If *Damon* or *Phrynus* discard prey when disturbed by the keeper, they do not ever attempt to locate and finish the meal. The uneaten portion will rot on the cage floor and eventually turn into a pile of mold, be consumed by the clean up crew, or just dry out. Spermatophores have some unique attraction as they are always consumed down to the bark in less than 24 hours, sometimes less than 24 minutes, when left in the cage (usually by the male). I have offered banana and beetle jellies with protein in limited trials and observed no feeding. Use of scavenging as a feeding methodology may be like trying to keep your dog healthy by giving it candy wrappers to lick. Nevertheless, Hodge (pers. comm.) reports once they grow hungry enough, adult groups of *Phrynichus exopthalmus* can be "trained" to feed on recently killed crickets and katydids in captivity. Initially there is little to no response but over time they eventually can become accustomed to this fare. A similar function of starvation and learning might be possible for other species.

Their appetite varies greatly—sometimes whipspiders will eat twice a day and sometimes they will not feed for weeks. Many hours are usually required for a single large prey to be processed into a small ball of discarded chitin. I have acquired specimens that ate three prey items the very first night and then did not feed again for nearly a month. After the first or second postultimate molt, some males do not even eat once a month. Females being conditioned for mating and egg laying should be fed a few times a week and may eat an incredible amount, but will return

Healthy immatures of all species will have rounded opisthosomas after a few weeks of feeding following each molt.

to feeding less often when carrying a broodsac. They tend to feed ravenously for weeks or months just after a successful mating. Adult females do not eat for a week or two following molts and after broodsac extrusion, but otherwise have a hearty appetite much of the time. The 2nd instars will eat almost every day and should be fed at minimum once a week. Optimally fed immatures will have huge abdomens that dwarf the prosoma and look like little balloons. Proper feeding will allow for rapid development. After a few molts they may be able to survive months without food but should be offered an appropriately sized prey item at least once every three weeks. If crickets are used it is important to remove uneaten prey within a few days, because an imminent molt will result in soft amblypygids being chewed on and killed by crickets. Amblypygids usually do not overeat, so food can be offered a few times a week as long as the previous food was taken. I have seen a few large females with abdomens that sag at an angle when fed excessively but I have not yet seen this lead to problems. (Risk due to damage from dropping during handling would be much greater.)

Unlike many pets, amblypygids can go a few weeks between feedings with no ill effects and may even be healthier when not fed too often. Heavily fed whipspiders are more likely to sustain damage from a fall. Additionally, if the desire is to improve longevity, feeding just a moderate amount can cause them to live much longer only inasmuch as it delays development time. However, whipspiders can starve to death. It is important to make sure the animals feed from time to time and prey is not just hiding or dying. Crickets are often used despite some risk because starvation is deadly. When feeding cockroaches it can be difficult to determine if the whipspiders are simply not hungry, uninterested in the prey, or if the prey has learned to avoid detection. With crickets the opposite is often true, with prey being killed even when the whipspider is not hungry. Overfeeding may not harm them directly, but they will eat only half of prey when they are not very hungry and sometimes just kill prey that is annoying. Rotting prey parts can directly poison the habitat (which is why isopods and springtails are recommended) or result in proliferation of harmful pest mites or molds that are capable of infecting the amblypygid. (Molds that feed on wood or substrate are rarely, if ever, harmful.)

Gut loading and dusting are methods that can be used to improve the nutritional quality of the prey item. There have been no controlled experiments but better nutrition should improve growth and possibly lifespan inasmuch as it leads to healthier animals (unless it leads to stronger, more aggressive animals that fight more and are capable of inflicting greater damage to cage mates or prospective breeding partners). Both methods can be difficult with amblypygids since prey must be eaten within a short time to prevent the external or internal additives from being lost to the cage floor.

Various powders containing vitamins and minerals are available and designed for dusting prey of insectivorous lizards. I dusted most prey offered to *D. diadema* for a few years and there seemed to be some improvement in speed of growth, but getting enough of the proper size prey can take a bit of time without adding an extra step and some prey are not easily dusted. (Prey are collected in a large deli-cup, a pinch of dust added and swirled to coat—not so easy with rapidly climbing prey in quantity.) I began dusting prey for the second *H. batesii* generation but also encountered a new, useful prey type (*Nauphoeta*) and used other prey more regularly (*Dorylaea* and *Eurycotis*). Additionally, the previously used prey cultures fluctuate in productivity and use, so any results would more likely reflect huge shifts in the overall diet.

Small immatures feed on prey their own mass with ease, 3rd instar *Heterophrynus batesii.*

Gut loading involves getting the prey item to ingest high protein dog food, powdered milk, pollen, or other nutritive item prior to feeding the arachnid, which then eats the prey and, hopefully, at least partly digests its gut content. Gut loading is more time and labor intensive and less easily observed. It is not always easy to see what prey have ingested or already defecated (let alone what portion the predator drops, refuses to ingest, or defecates). Gut loading is often tried with captive-bred flies since they have little nutritional value of their own, but its function is primarily untested. Gut loading may also reference heavy fortification of media used in rearing prey, though the actual gut contents may be minor. Rayor (pers. comm.) reports having good success using the commercial fruit fly diet from Carolina Bio, adding several large pinches (2 teaspoons or so) of nutritional yeast to the media, then adding regular bread yeast to the top when rearing *Drosophila hydei* fed to 2nd instar *Damon diadema*. (The *Damon* are offered well-fed house flies as soon as they can handle them at 2nd instar, while crickets are only included after a molt or two. The house flies are leftovers from a toxicology lab and are a strain with deformed wings and bad eyes. House flies are reared on a combo of wheat bran, hardwood chips, commercial calf manna diet, brewers yeast, and a few other nutrients. A small portion of the *D. diadema* diet also includes cabbage looper moths raised on pinto beans.) Keep in mind, without specific precautions (and according to species even with them) amblypygids fed only fruit flies can die even if they ever make it to a molt. Gut loading experiments using flies as a singular prey should only be performed if the premature loss of every animal is not a concern.

The hobbyist who keeps more than a few dozen specimens or wishes to rear up captive broods will almost certainly want to culture at least a portion of the necessary prey items. Rearing prey requires an investment in time and space, but can save a lot of time and money in the long run. It can be very difficult to get small food items of the appropriate size from local shops to feed immatures. The first step is to determine which prey will be cultured. A number of common feeder insects have been tried and are essentially useless as food for amblypygids because they are almost never eaten—these include mealworms, pea weevils, red flour beetles, and springtails. Despite the presence of isopods in most cages I have not yet noticed one being stalked or consumed. Waxworms reared to maturity and small grain moths may be used but are not functional as a standard feeder because timing and production generally do not coincide with need.

The following prey items are commonly used, readily reared, and proven foods.

Cockroaches

The cockroaches are a huge group of more than 4,000 species that vary greatly in natural habitat and modes of reproduction, and only half a dozen of these have become infamous for making a living in human structures. The vast majority of roach species are not pests. Exotics, primarily livebearers, are normally kept because they cannot infest homes. Functional types are listed below, loosely from most to least useful.

Schultesia lampyridiformis Firefly Cockroaches.

1. Firefly cockroach *Schultesia lampyridiformis* is a live-bearing, high profile species that does not burrow and is nearly perfect in size for all uses. This is the single best feeder for whipspiders. Newborn fireflies are small enough to feed to almost any 2nd instar, though adults are not as big as a large cricket. It is relatively fast growing, but it would take some time to establish a large colony to feed dozens of whipspiders.

2. Colorful, high profile egg-layers like *Dorylaea orini* and *Eurycotis decipiens* make excellent feeders if available. (*Eurycotis floridana* and *Eurycotis opaca* hatch out too large for small whipspiders.) They are excellent prey for amblypygids since they do not burrow and the warning coloration and innocuous defensive exudates do not deter feeding. The homely *Blatta lateralis* and various *Periplaneta,* among other relatively similar species, are probably as functional and much less expensive to start up, but are among the very rare cockroaches that might survive outside their captive enclosures.

3. *Nauphoeta cinerea* do not burrow in fine, compacted substrate, but they can hide under rather tiny chunks of loose substrate, maintain a low profile, and may avoid being eaten indefinitely. They can be useful, but it is not rare to find every prey item offered over the preceding month hiding under a decoration. The only whipspider I have seen take these out at a good clip is *H. batesii*, though they are also eaten by *P. barbadensis* and uncommonly by *P. mexicanus* and *D. diadema*. Pallid cockroaches, *Phoetalia pallida*, are much smaller and better at escape, but similar in behavior. Early instar hissing cockroaches

Dorylaea orini.

Eurycotis decipiens.

Paraphrynus mexicanus feeding on an adult Pallid Cockroach, *Phoetalia pallida*.

from the genus *Gromphadorhina* and the tiny *Oxyhaloa deusta* also fit here.

4. Ornate velvets, *Deropeltis paulinoi*, and the related *Deropeltis* sp. "jinka" are extremely productive and are eaten at times, but produce toxic quinones. These are pretty much a last resort feeder because the amblypygids quickly learn to avoid them, but they are eaten anyway if the whipspiders are extremely hungry (so still far more useful than burrowing prey). The viviparous *Diploptera punctata* is similarly equipped with quinones but starts out larger and breeds more slowly, making it even less useful.

5. *Panchlora nivea* nymphs burrow but the adults often stay topside and can be a good treat for whipspiders large enough to eat them. *Pycnoscelus* spp. nymphs are similar but even adults are generally useless as prey.

6. Large live-bearing, burrowing species like Discoids, *Blaberus discoidalis*, and Dubias, *Blaptica dubia*, may be of some use but normally go straight into the substrate never to be heard from again. These are usually the least expensive, while the more expensive larger pet species *Blaberus, Archimandrita, Eublaberus*, are similar in function. Even Discoids grow large, so a half grown nymph would be a huge meal for the largest female whipspider, but they burrow and generally are not eaten. It is possible to trick amblypygids into eating these, but I have offered *Blaberus* nymphs of varying sizes to aggressively feeding *Heterophrynus* with almost no response. I only chanced upon one being consumed three weeks after a dozen tiny *Blaberus craniifer* nymphs were placed in the habitat.

Pros: Cockroaches do not attack molting or sickly arachnids. (Even the somewhat predatory *Eublaberus posticus* is unlikely to present a problem unless the habitat is teeming with them.) Many, if not all, cockroach species are very nutritious according to husbandry results. They do not jump and do not chirp. A number of species cannot climb glass or other smooth surface. They can survive weeks or months in the cage until the whipspider is hungry. A decent colony produces a far greater volume than crickets in the same amount of space. Most keepers find feeder species easy to culture. Upkeep requires very little time or expense.

Deropeltis paulinoi: not a very useful feeder.

Cons: The most useful species are primarily sold as culture starts, so it is necessary to maintain a colony, which may only be worthwhile for those with quite a few animals to feed. Some are expensive. Colonies can take months or longer before excess can be harvested for use as prey.

Rearing Instructions: Roaches are often kept in five-gallon buckets, plastic totes, or fifty-gallon trashcans. It is important to have some screen ventilation or they die. Most species breed well around 75° F (~24° C) though higher temperatures can speed up production. Egg cartons or pieces of rough wood or bark are placed in the cage for climbing and hiding. A water dish containing gravel or sponge to prevent drowning should be provided or an area of damp substrate. The primary diet consists of a kibbled pet food such as dry cat food or monkey chow; dry dog food is the most commonly used. Fruit, primarily apple or pear cores, should be offered from time to time to the live-bearing genera (i.e. *Panchlora, Blaberus, Eublaberus, Schultesia, Nauphoeta*, etc.) to improve production. Egg-laying genera (including *Dorylaea* and *Eurycotis*) should be provided cork bark slabs that are misted from time to time. The soft wood provides a deposition site and misting keeps oothecae from drying out. Cages that contain glass-climbing species should have the upper two inches coated with petroleum jelly to help prevent escapes when the lid is removed.

Acheta domesticus House Crickets (hypermelanistic variant).

House Crickets

The house cricket (*Acheta domesticus*) is the most common feeder. The stock pictured are representatives of a hypermelanistic (black) form of the normal feeder *A. domesticus* cultured to hatch out at lower temperatures, that has been used to feed whipspiders for years. They are not field crickets (*Gryllus* spp.), which have short rear wings, black nymphs, and are much more difficult to rear.

Pros: These are active, time-tested prey items that jump around to attract attention and do not burrow or hide in cracks to escape. A healthy whipspider will never starve if offered appropriately sized crickets. Feeder crickets can be purchased at nearly any pet shop and through the mail. They are commonly available in numerous sizes to accommodate the size of the amblypygids being fed.

Cons: Crickets hide under objects and are highly aggressive. Uneaten crickets are extremely dangerous, especially to molting animals. If not eaten right away they can die, stink terribly, and serve as food for mite outbreaks. (As luck would have it they never seem to die if the whipspider is about to molt.) Except for bulk purchases, crickets can be very expensive to purchase over the long run. Rearing large quantities is space and time intensive and not always easy. Crickets are excellent jumpers and often escape into the home where they can chew on books, food, and other personal items. The sound of a loud male chirping away every night interrupts the sleep of many hobbyists' spouses. Small crickets, especially pinheads, are very difficult to handle and those that do not jump away are often squashed by the act of picking them up. Sometimes they are unavailable in hot summer months, since they do not ship live at high temperatures.

Rearing Instructions: Usually at least two ten-gallon terrariums are needed to rear crickets. A few large pieces of egg carton are added for molting and hiding. Adult female House Crickets lay a few hundred eggs in any potting soil or dirt that is an inch (25 mm) or more deep. Removable trays can be relocated to new cages if varying sizes need to be isolated to feed different sized whipspider broods. The dirt should be kept moist until tiny pinhead crickets hatch out after a few weeks. The nymphs and adults do well on many foods such as leftovers, old fruits and vegetables, and dry dog or cat food. Water in a sponge or gravel filled dish should also be provided to the growing crickets. It is necessary to use a screen or otherwise ventilated cage or the crickets will die within a few hours

Working on a cricket, *Phrynus barbadensis*.

to a few days. An area or side of the soil should be kept constantly moist to keep the eggs from drying up. Additionally, at least a 60-watt incandescent light bulb located directly atop the cage is a necessity if the ambient temperature is below 75° F (24° C), to provide additional heat. Without this source of heat it is often impossible to rear a single cricket because the eggs will not hatch. After two to four weeks, hundreds of tiny hatchlings emerge.

Fruit Flies

Drosophila melanogaster is the most commonly cultured species and is available in flightless and wingless mutations. These are a useful and valuable filler but horrific singular prey. *Drosophila hydei* is a larger, sometimes less easily maintained species, also available in various strains such as flightless and gold.

Pros: Fruit flies are very useful for getting tiny immature amblypygids interested in feeding. Early instar *Damon* and *Phrynus* can be kept well fed on *D. melanogaster* since it is difficult to produce enough crickets of small enough size for large or multiple broods. Fruit flies have a relatively rapid turnaround time, can be produced in quantity, take little space, and are readily available as established cultures.

Cons: Drosophila is deficient in amino acids and requires significant precautions for use as a primary food, and even then may only prove useful for certain hardy species. Use as a singular food is a well-known cause of "curly leg syndrome" and crippling molts in tarantulas (McMonigle 2006) and usually cannot sustain the growth of immature centipedes, scorpions, nor amblypygids through a single molt. Fruit flies can be used successfully as a singular prey for mantids, orb-weavers, and jumping spiders, but none of these are amblypygids. Hebets (2002) noted wild amblypygids in Costa Rica fed on various items, mostly crickets and katydids, but also cockroaches, spiders, harvestmen, moths, and even diplopods, but never a single fly. Whipspiders greater than 1/4" (5 mm) body length tend to ignore the tiny flies and *Drosophila* appear to be of no value in feeding *Heterophrynus* which refuse to catch them. Flies die after a few days inside the cage and should be given a tiny apple slice to keep them alive. Cultures have a strong fermenting odor and new ones must be set up at a regular pace as long as a supply is needed.

Rearing Instructions: Fruit flies are kept in a sealed container with a breathable lid and one to two inches of media in the bottom. The lid keeps in *Drosophila* and keeps out phorid flies. Tall 32 oz (1 L) deli-cups with the coffee filter built into the lid are the most common culture container. The media can be made of numerous items including oatmeal, banana, apple, etc. but

Fruit flies, wingless *Drosophila melanogaster*.

the easiest and cheapest consists of instant potato flakes and water. (Various media can be different—Gerber mixed grain cereal produces flies three days faster, but cultures burn out too quickly.) The exact ratio depends on ambient humidity, but about 1/2 cup (0.25 L) potato flakes and slightly more than 1/2 cup (0.25 L) water for one 32 oz (1 L) container. (Vitamins, protein isolate powders, and nutritional yeast can be mixed into the media to improve food value, though very little may transfer to the flies.) The potato should stiffen up within a minute of being mixed or more flakes may be needed. If media is too watery it will drown the adult flies before they lay eggs and the media will pour out when the container is tilted for feeding. A pinch of yeast is sprinkled on the surface to stimulate egg laying, deter mold, and improve success and growth. A dash of liquor, vinegar, or apple slice will work almost as well. It is important to keep cultures in series because they usually exhaust after a few weeks even if all the food is not used up. With the large containers, setting up a new one every two to three weeks provides continuous flies while with little vials, half a dozen had to be started each week. Remove about three dozen of the first wave of flies to start a new culture. If a culture becomes overrun with mites, it has been kept around too long. Larvae will eat slime bacteria and mold that starts to grow on the surface about a week after the culture is started. When the first pupa is seen on the side, throw in a crumpled paper towel so pupae can form on it or most will drown and die. If added too early, the paper can reduce or prevent egg laying. (Excelsior, which is long, very thin strips of aspen wood shavings, or corrugated cardboard tubes can be placed earlier since they do not adversely affect egg laying.) Mold inhibitor is not necessary and may be impossible to acquire these days with increased regulation (except as part of commercially available media mixes). Mold only overgrows cultures that were started with too few flies, not provided an egg-laying stimulus, or made too dry. The temperature should be from 75°-85° F (24°-29° C). Warm temperatures shorten the cycle time by a few days, while too cold of conditions can stop production.

8
SPECIES LIST

Acanthophrynus coronatus (Butler, 1873) A gravid female may reach nearly 2″ (45 mm) in body length, whereas the largest *Damon* and *Heterophrynus* are generally from 30-40 mm. This is the largest known whipspider in the world according to body length, although it does not hold the record legspan, whipspan, or palpspan. The ambulatory legspan of a large individual is reportedly from seven to nine inches unverified. Were the maximum true it would still be 2.5″ (64 mm) shy of the largest *Heterophrynus* species, but there may be limited room for debate since the ambulatory legspans of these species are not well documented. On the other hand, the whipspan and palpspan are not even remotely close to those documented for some *Heterophrynus*.

Acanthophrynus coronatus is the only member of the genus, but it is a member of the subfamily Phryninae with the genera *Paraphrynus* and *Phrynus*. The most notable identifying feature of this species is a fringe of small spines across the front of the prosoma—this "crown" explains the specific name. (However, if you look at the close-up images of *P. marginemaculatus* and *P. barbadensis* included in this text, they have a similar though less pronounced row of small spines.) The chelicerae each have a row of flattened spines on the inner surface used in stridulation, but this would be very difficult to see on a living specimen when not heard. The palps of the adult are orange-brown and the basitarsus has just one large dorsal and one large ventral spine, whereas related species have additional spines. *Acanthophrynus* is the only whipspider I know of with relatively long palps that does not display obvious sexual dimorphism of the palps (males might average smaller or have a more deeply scalloped basal prosomal margin). Quintero (1979) references the lack of dimorphism in the *A. coronatus* pedipalp when describing sexually dimorphic palps for some *Paraphrynus* species. The smaller male pictured appears to have longer whips than the female, despite his much smaller size.

Acanthophrynus is unheard of in US collections but is listed because it is, or was, reportedly found in the United States. The type specimen locality is listed from USA, California (Harvey 2003). There is a single fuzzy photo of a specimen that has palp spines, leg dimensions, prosomal shape, and abdominal tergites almost certainly identifying it as *A. coronatus* that was found in Nogales in Santa Cruz County, Arizona, posted on bugguide.net. (All the other specimens depicted from Arizona on

Acanthophrynus coronatus pair in Jalisco, Mexico. © Eddy Hijmensen

This is the male feeding on the cricket it just caught. © Eddy Hijmensen

that website are easily identified as *Paraphrynus mexicanus*.)

In the Wild: "This photograph of a wild pair on a tree in Mexico was an unusual find. While looking for *Brachypelma klaasi* in the Chamela reserve in Jalisco we came across this pair. As we did not have our cameras on us, we had to walk back to the car to get them. When we returned the pair was still in place on the tree, nicely on eye level. They did not hide from our headlights and were not bothered by the camera flash. Even better, a cricket landed on the tree close to the male. In a split second he lashed out and caught it. Specimens are usually seen on trees or on rocks at night. During the day they hide in crevices between rocks. Apart from the pair on the tree, we only found single specimens. We have encountered this species in the Jalisco dry forest and Balsas dry forest ecoregions" (Hijmensen pers. comm.).

Distribution: Mexico, USA—California and Arizona (Harvey 2003).

Charon grayi (Gervais, 1842)
FILIPINO WHIPSPIDER

This species is only listed to represent the family Charontidae, as specimens have never been seen in the North American hobby. It is among the larger whipspiders, with a body size reportedly comparable to *Damon medius* and a whipspan that may reach almost 20″ (<50 cm). It was described from the Philippines, with populations from various Southeast Asian areas later ascribed to it. This species has been kept by some European enthusiasts, but reports of successful breeding are scarce, if not altogether lacking. In nature, it is found almost exclusively in natural or artificial caves and on large rocks in rainforests, but only rarely on adjacent trees. Specimens are spaced apart by some meters and quickly retreat to the nearest crevice if disturbed. They probably feed mainly on cave crickets, which live in the same habitat (Schwarz pers. comm.). Males have longer palps than females, although the difference is likely not as apparent on young adults.

Distribution: Philippines, while populations from Indonesia, Malaysia, Singapore, Papua New Guinea, Solomon Islands, and Palau may not be conspecific (Harvey and West 1998).

Damon diadema (Simon, 1876)
GIANT TANZANIAN WHIPSPIDER

Damon diadema, commonly imported from Tanzania, Africa, is among the world's largest, most impressive, and certainly most famous whipspiders. The huge pedipalps tear apart crickets, but seem apparently unable to scratch human skin. *Damon* is an archaic spelling of the word demon. The species name comes from the word diadem—the king's crown—so the scientific name could be translated as "the demon king." However, *diadema* probably refers to the spines on the palps or markings on the prosoma. The legspan of mature adults is around 5″ (ca. 127 mm) and the whipspan nearly 12″ (<300 mm). At an inch and a half body length, this is the largest species not part of the western *D. medius* generic subgroup once known as "Titanodamon." Mature *Damon* vary in size and a large adult can be more than twice the size of a small one. They mature in two to three years and are sexually mature well before reaching full size. Older males have much longer pedipalps but this is not the case with younger animals. The difference can be difficult to see on specimens even as they

Adult *Charon grayi*, on tree bark, Panay Island, Philippines. © Christian Schwarz

Charon grayi, juvenile, Panay Is. © Christian Schwarz

approach maturity. Size is gained through post maturity molts, so the largest individuals are often the oldest. However, only the first postultimate molt results in significant growth, the second results in a much less impressive increase, and all subsequent molts do not result in growth. Adults live many years but are unlikely to reach the hypothetical 25-year mark suggested in some writings. My experience with multiple breeding adults is most live about five years from 1st instar.

Many of the *Damon diadema* imported from Tanzania have been incorrectly labeled as *Damon variegatus*, but so far not a single import in North America has proven to be *D. variegatus*—though there has been the very rare *Damon medius* and *Phrynichus* spp. labeled as *D. variegatus*, as well. Even in Europe the vast majority of specimens are *D. diadema*, as the misidentified photographs on the net and in Beccaloni's (2009) arachnid book can attest. *Damon variegatus* is smaller, has short palps each with one rather than two enlarged spines located ventrally on the trochanter, is often darker colored, and females possess huge pleural membrane folds. These cover a large portion of the surface of the broodsac during gestation. Indications of the huge pleural folds can be visible on the ventral margins of the female's opisthosoma even prior to maturity. *Damon diadema* pleural folds cannot be seen in the slightest on females lacking a broodsac and when in use cover only the outer edge of the broodsac like many *Phrynus*. Some older range maps show *Damon variegatus* is not found in Tanzania, but some papers claim it is found in the extreme west—far from exporter access. *Phrynichus* mislabeled as *D. variegatus* are even easier to differentiate as they have fewer large spines on the tibia of the palp and males have much narrower palps, which gives the adults a very different, less menacing appearance. *Phrynichus* also lack leaf-like setae on the tarsi of the antenniform legs as these are only found on *Damon* species.

Adult female *Damon diadema*.

Damon diadema accept a wide range of humidity levels and can survive weeks without moisture that would kill many other amblypygids. Nevertheless, available moisture is necessary for long-term survival and successful captive reproduction. While moisture is extremely important, unrelenting high humidity during gestation can be problematic. Captive-bred are hardy and long-lived under adequate conditions but some imports survive only weeks after arrival.

My Quest to Acquire the Demon King: For more than a decade this has been the only large, regularly available amblypygid. Specimens are usually wild-caught from Tanzania but captive hatched are reasonably common. This huge, beautiful beast has been captive hatched by many dozens or hundreds of enthusiasts, though captive

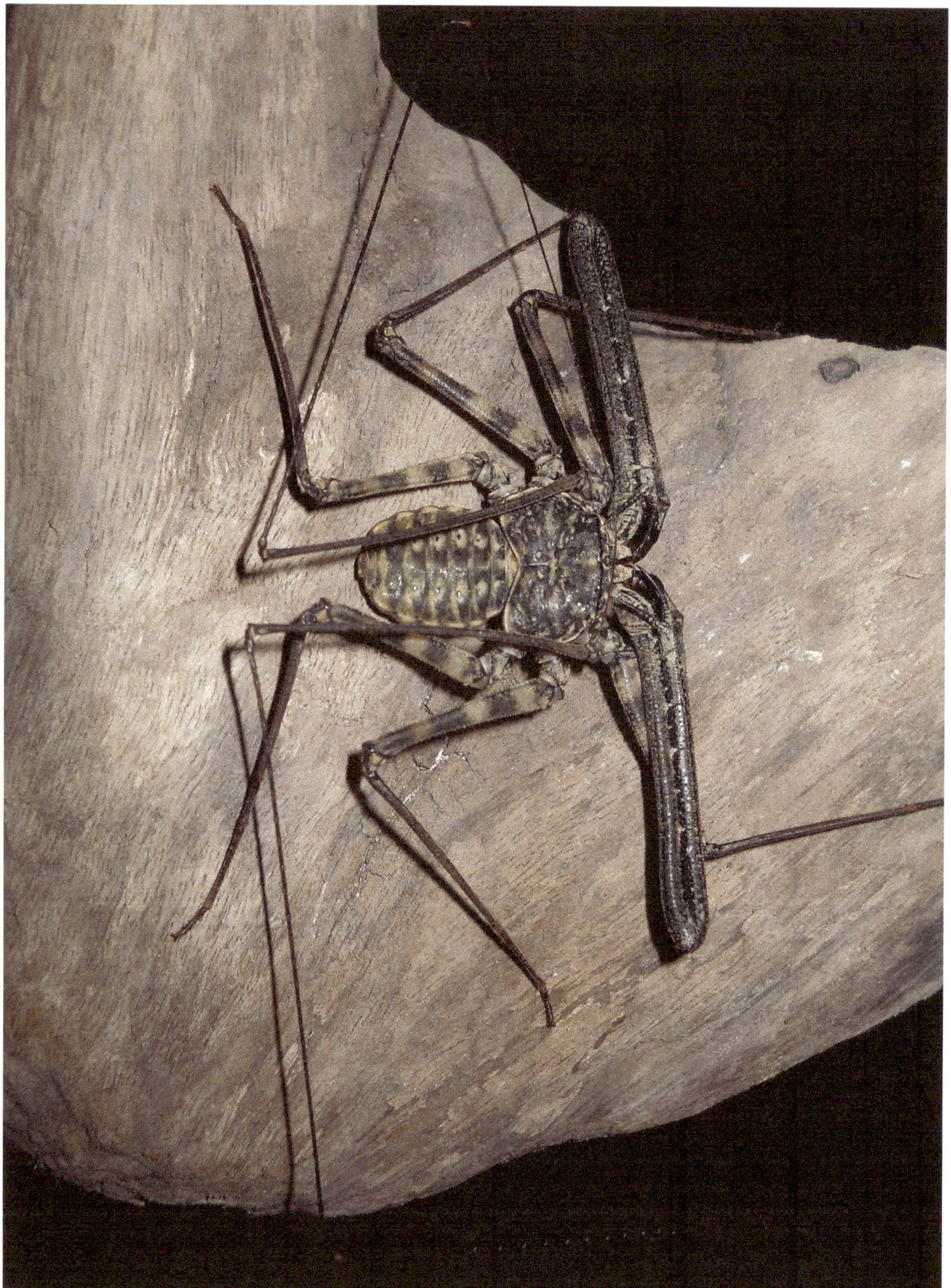

Older adult male *Damon diadema*, leaf-like setae on the whip can be seen as a row of white dots near bottom of image.

breeding through subsequent generations is not so common.

Damon of any kind had been near impossible to find until the beginning of the new millennium. After a long search I traded for some captive-hatched babies in 2000. (I had acquired two 2nd instar in 1994 that were fed fruit flies for six months and never molted. At the time I did not know whipspiders, and a number of other invertebrates, cannot develop if fruit flies are used as a singular food.) 2001 was the first time I saw adults for sale and sent money for a pair immediately. (It was certainly a pair because *D. diadema* is extremely easy to sex—the mature male's pedipalps folded are longer than the femur of the front leg, while the female's are shorter.) The male came in dead but the female seemed healthy. She died two weeks later.

In the summer of 2003 I received a dozen adults directly from an importer. They looked absolutely beautiful and healthy on arrival. I was beyond thrilled to find every female carrying a broodsac full of eggs! Two of the females were especially huge and had more than seventy eggs each. I wondered how I was going to feed and care for hundreds of baby amblypygids. The next day one female was dead, and all her eggs with her. Over the next three weeks every female and all but one male fell over dead. It was horribly anticlimactic.

Some hobbyists have had great luck getting healthy wild-caught adults, but my score is one win out of fifteen. I picked up a few more captive-hatched young in 2002, which matured in late 2005. With the conspicuous exception of husbandry mistakes, my survival for hundreds of captive-hatched young has involved no mysterious deaths. Survival through the early instars has been nearly perfect. After they are old enough to be separated, a small number are killed by crickets or, rarely, molting issues from slight desiccation, and once they grow large enough to be reintroduced, some are lost to cannibalism and aggression.

Reproduction and Generations: A surviving female from the original captive-hatched babies (2000) seemed to have been mature for a long time by winter of 2003. I figured it could not hurt to put her and a male together. (They were from the same breeder, different mother, same father, and had originally been traded to me as "*Phrynichus* species" from Tanzania.) He showed no interest in her for the eight months they cohabitated, but then she molted. Within weeks there were spermatophores glued here and there on the cork bark.

A few months later, at the end of August 2004, she was seen carrying a broodsac full of eggs. She continued to eat a cricket a week. Female *D. diadema* continue to have an appetite while holding eggs. A female might survive incubation without feeding (as suggested by some authors) but at best it might not kill her. Eggs take a long time to hatch and during gestation are internally nourished.

It was a long three months of gestation before I was certain the eggs were fertile. I could see notable development of the eggs through the semitransparent membrane as the postembryo is not a perfect sphere. She was moved to her own cage. The female can be kept with other adults until just before hatchlings molt and disperse. Other hungry adults in the cage would most likely eat the little guys and their prey can be dangerous as well. It was another month before her young hatched on December 21, 2004.

1st instars molting to 2nd instar.

The membrane began to swell forty-eight hours before hatching and began to crack eight hours before hatching. Three of the 1st instars did not make it to her back and hung out together in a depression in the dirt. The female does not feed on 1st instar immatures and provides no visible care. Since they seemed healthy and the humidity was high throughout the cage, I left them. For no apparent reason, on the morning of day three, two were dead. I carefully coaxed the survivor onto the mother's back where it temporarily became surprisingly mobile. It positioned itself among the others and remained in place until the first molt. Over the following days the white legs and prosomas of the 1st instars developed a progressively darker pink coloration and the opisthosomas grew darker green.

On January 1, 2005, 34 young molted to 2nd instar and dispersed throughout the cage. I had accidentally allowed the wet substrate to become slightly dry and three lost a leg or two in the molt. They were partly stuck in the old exoskeleton and had to be cut free.

The young were moved to a 1.5 gallon (5.7 L) bucket with an inch of wet substrate and three 5″ x 4″ pieces of egg carton. The entire center of the lid was cut out and covered with microscreen. They were fed tons of flightless *Drosophila hydei* and pinhead crickets until their abdomens swelled like little green balloons. It was surprising how much they ate.

On February 9, 2005, thirty-nine days following the dispersal molt, the first 2nd instar *Damon* underwent its second molt to 3rd instar. Two more molted on Feb. 13th.

The *Damon* broodsac is impossible to see unless viewed from below and females do not flash the abdomen unless you force them to for a photograph.

Damon diadema postembryos getting ready to hatch.

One or two continued to molt every few days at a regular pace until the second to last one molted on day seventy-four (March 16). The three damaged immatures were among the last ten to molt. One straggler did not have any damage or visible reason for molting late, and finally molted on day 119. The first to molt for the third time to 4th instar also molted on day 119. Three weeks after this molt there seemed to be signs of aggression and all were separated. On day 187 they began to molt for the fourth time to 5th instar.

Rearing containers for young consisted of 16 oz (0.5 L) deli cups. Tall, 32 oz (1 L) or larger deli cups would be used later for a sub-adult or adult. A piece of metal window screen is glued to the underside of the lid to provide a molting surface. A section of plastic (used for yarn crafts and available at any craft store) is cut the length necessary to reach from the bottom corner to the top, at an angle. This plastic is important and used as a ladder, since the whipspider cannot climb smooth sides of a container. A 1"-2" (2.5-5 cm) square is cut into the side and a piece of microscreen or cloth is glued over the hole. Larger screens can be used but increase problems with drying out.

This 5th instar *Damon diadema* is closer to adult coloration.

L2 *Damon diadema* are quite colorful.

Once the cage is built, a 1/4" (6 mm) layer of coco fiber or soil is placed on the bottom for moisture retention. The rare deaths post-separation were exclusively due to hungry crickets. Since the whips of older animals can be longer than far larger enclosures, I have had a very few over the years sneak in the tips under a closing lid during feeding. This is only a problem if it goes unnoticed because after a day or so the trapped whip will be autotomized down to the patella.

The female molted 189 days after the dispersal molt of her young and increased more than 30% in size. The male molted fifteen months after the last molt (just before his offspring hatched). I was impressed to see he had gained 20% in size. While still green, he jumped off my hands and fell four feet to the cement. He appeared unaffected but died the next day.

I was able to acquire another adult male to be placed with the adult female. They coexisted "happily" in a small cage. He was much smaller than other mature males I had previously, but he did create a few spermatophores (and fertilized eggs that hatched in Nov. 2006). Months later a new female was placed with the pair. She seemed to be an outcast so I only left her in for a few days at a time. I had not considered she would be killed since she was not near a molt. On day three of the fourth attempt she was found on the substrate and was very sluggish. She had no missing parts but on close inspection her prosoma had two, shallow, 1/2" (12 mm) long lacerations—evidenced by a trace amount of dried blood. She was isolated immediately but died after a few days.

Multiple adults were reared and siblings were not mated, though all were related offspring of the original groups. The subsequent generation (her grandchildren) hatched out in late 2007 and early 2008 and the next (great grandchildren) in late 2010, but there were related concurrent broods in the off years. The original F1 generation was the slowest growing but the subsequent generations matured in just over two years and produced young before the end of year three. Deaths due to crickets were virtually eliminated post-separation due to increased familiarity with signs of impending molts. Improved growth was a result of becoming more accustomed to the feeding and husbandry needs. With close attention and constant feeding it is possible for young *Damon* to molt every few months, though molts still grow progressively spaced out as they grow larger. *Damon diadema* usually become sexually mature following the seventh molt, but females may not mate, and often do not produce broodsacs even if mating was attempted, until after the eighth molt. Adults usually molt once a year. However, after a specimen has reached maximum size (following two postultimate molts) it may not molt for as long as thirty months.

Broods: In my care the usual number produced by captive reared specimens was 30-40, with only the older, larger females producing around fifty. Dixey (2002) reported 56 and Cernicka (2003) reported 54, both from wild-caught adult females mated in captivity. Large, wild females have arrived with greater than 70 eggs in one broodsac. Gestation of the broodsac takes approximately four months post-extrusion. 1st instars molt ten to eleven days after hatching.

Distribution: Kenya, Tanzania, and Yemen (Weygoldt 2000).

Damon variegatus photos like this are hard to come by (unless the photos are *D. diadema*). © Michael Seiter

Damon medius (Herbst in Lichtenstein and Herbst, 1797)

This handsome creature began showing up on vendor lists in late 2007 but had been imported in extremely limited numbers years earlier. It appears similar to *D. diadema* but can grow larger, has more brown

Compared to a hand, *Damon medius* female © Wade Harrell

in the body coloration, and is outfitted with more elongate, spinier pedipalps. The old adult female can be mistaken for a male (if there is no male to compare her with) because the elbow of the palps extend further than the front femurs, unlike *D. diadema*. It is easy to sex specimens because the adult male's pedipalps are visibly longer and the genital operculum is sexually dimorphic. Females have arrived gravid and there were captive-hatched immatures for sale, but I have not seen any sign of this species on vendor lists since small C.H. were available in 2008. This and other members of the West African *D. medius* group, like *D. johnstoni*, were at one time separated into the genus "Titanodamon."

Damon medius female © Wade Harrell

Although some of the smallest whip-spiders live deep inside the nests of eusocial insects, some of the largest may be found in the periphery. Some termite mounds can be extremely large, so much so that *Damon medius* are commonly observed in the ventilation openings of savanna termite mounds in the Ivory Coast (Schwarz pers. comm.).

Broods: +/- 40

Distribution: While this species ranges over western Africa, the import label is "Ghana West Africa."

Heterophrynus Generic and Specific Identification Features

Heterophrynus have the greatest legspan of any fully terrestrial arthropod, but the specialized front legs are of little value in identification. The antenniform front legs are in fact legs, and a 23.6" (60 cm) whip-span is documented for specimens of *H. batesii* and *H. longicornis*. (The largest specimen of the largest spider does not reach 12" in legspan.) Remember, this is the measurement of normal, large specimens, not the single largest specimen ever recorded as seen in other arachnid measurements (try finding an 11" *Theraphosa* or 9" *Hadogenes*). Even the insects cannot compare to *Heterophrynus*—the world's longest insect specimen, confirmed by British scientists, a particularly large representative of a walkingstick species, has a legspan documented at only 22" (56 cm) (Romero 2008). Other than sheer size, the whips play no part in identification since the span and articulation are somewhat variable. Counting the articles and tiny sensory hairs according to type and location under magnification could take days, and specific count locations are documented across multiple specimens of few,

Heterophrynus alces with beautiful purple catch basket,
a few days before the second molt in captivity.

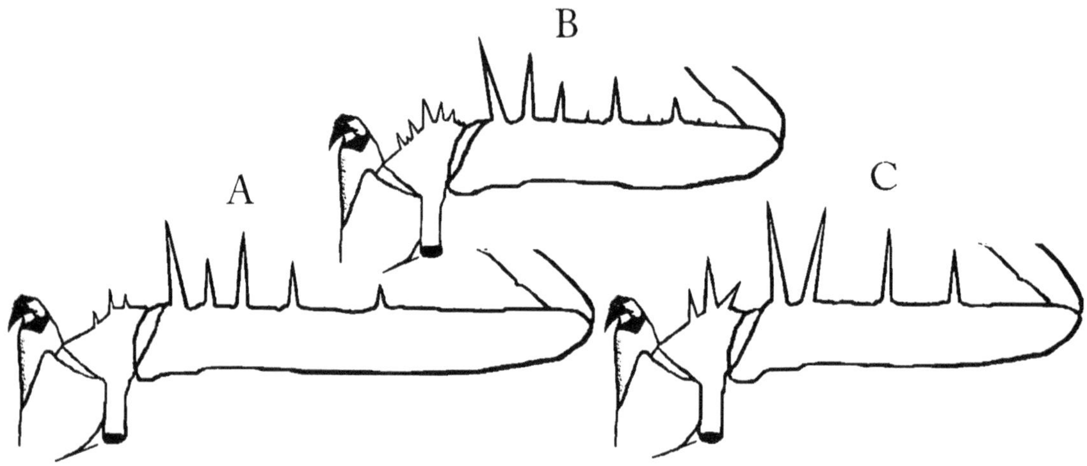

Heterophrynus pedipalp femur diagrams.

if any, species. Additionally, regenerated whips can be much shorter and differently articulated. Furthermore, whips in this genus can be sexually dimorphic as, at least for the *H. batesii* pictured, the male's whip is 20% longer than the unregenerated whip of the female of the same adult stage.

Heterophrynus are easy to differentiate from other genera because there is a pronounced, backward facing, ventral apophysis on the palp trochanter, and the dorsal palp tibia (catching basket) is outfitted with three very large spines of similar length. The species identifications are based primarily on the spine structure of the ventral side of the palpal femur, augmented by female genitalia (male genitalia are highly variable).

The living species from Peru has the first three large spines on the ventral surface of the palpal femur that, from the base, are long, short, and then long (fig. A). It is differentiated from the only other species with similar spine arrangement by the smaller, deeply hooked genital sclerites (a pair of hardened plates connected to the spermathecae) of the female and the far greater relative length of the male palps. The palps are long and the palpal femur is greater than twice the width of the prosoma, even in very small male *H. batesii*. The other species with fig. A spines, *Heterophrynus longicornis,* has relatively short palps with palpal femurs less than 1.6 times the width of the prosoma even in the largest male, and the female's genital sclerites are long and minimally curved (Weygoldt 2002).

The living species from Guyana has pedipalp ventral femurs with many small spinelets between the main spines, while the third large spine from the base of the pedipalp femur is shorter than the second and fourth (fig. B), which together identifies it as *Heterophrynus alces*. The only other species with large spines similar to fig. B is *Heterophrynus armiger*, but without the large number of small spines (Weygoldt 2002).

The dried specimens from Peru have the first four large spines of the ventral palpal femur descending in size (fig. C) and the palpal femur of the male is barely longer than the carapace, which identifies it as *Heterophrynus elaphus*.

Heterophrynus alces Pocock, 1902
DEIMATIC GARGANTUAN WHIPSPIDER

The spectacular members of this genus are endemic to forests of the Amazonian region and may only have opportunity for continued survival in captivity after most of the rainforests are cut down. They live in caverns among the buttresses, aerial roots, and rotten cavities of massive tree trunks, and without old growth they have nowhere to live. Young trees might provide adequate hunting ground but not daytime cover.

Seeing any *Heterophrynus* spp. in its native habitat must be an amazing sight. I have heard similar stories relayed in awe from half a dozen people who have visited South America (probably everyone I have talked to who has been there) about the time they ran into one or more of these amazing creatures. When I first saw a live *H. alces* up close, I was awestruck by its size and attitude. *Heterophrynus* look far more unreal and intimidating in three-dimensional movement rather than pinned flat to a board. I imagine it would be far more amazing to see on its own turf. Although it is among the smaller representatives of its genus and has a smaller body than large *D. diadema,* its longer legs and three-dimensional attitude make it seem relatively gargantuan.

There are at least eleven *Heterophrynus* species. All of them have spectacular pedipalps outfitted with huge spines, but only a couple have incredibly long pedipalps, such as *H. cheiracanthus* and *H. batesii*. (*H. longimanus*, which means "long hand," is a junior synonym of *H. batesii* (Giupponi 2002).) This handsome, short-palped species was sold as one of the larger species believed by many to have elongate pedipalps—and labeled *H. longicornis*—but this species and *H. longicornis* have short palps. *Heterophrynus brevimanus*, which means "short hand," is a junior synonym

Heterophrynus alces 2nd molt in captivity.

of *H. longicornis* (Giupponi 2002), so the name *longicornis* meaning "long horn" must certainly reference the long antenniform legs (just as longhorn beetle and longhorn grasshopper are references to antennae), not the short palps. Still, the palpal femur of this species was only as long as the prosoma width, even shorter than that of *H. longicornis*, which is never longer than 1.6 times the width of the prosoma, and the femoral spine structure was all wrong for that species. These were not simply immature specimens, because one of the pictured specimens has molted three times in captivity without gaining appreciable size, a certain sign of maturity. Even the females of species with long palps have long palps, although they are shorter than the male's. Still, *Heterophrynus* species

The *Heterophrynus alces* male that appeared dead on arrival.

with shorter palps can make up for it with larger spines on the capture basket. Although my dream species was always *H. batesii*, after so many years of searching and hoping, seeing any *Heterophrynus* alive was a dream come true. This is the smallest of three *Heterophrynus* species I have seen specimens of, but it still has a walking leg span in excess of 7″ (178 mm). The short palps set it apart from the species discussed next and the huge basket of capture spines make it an incredibly beautiful species. It also displays spectacular and unique defensive behavior.

The commonly available *D. diadema* and *P. marginemaculatus* display two-dimensional movement of the unfolded pedipalps (opening and closing with little more articulation than many arthropod jaws). Since I was used to limited up and down movement, the versatility of this species was initially surprising. *Heterophrynus* can hold the palps backwards, up over the prosoma for downward and forward capture movement like some spider chelicerae. The palps can also be opened from side to side or combine both vertical and horizontal movements in a single swing.

Heterophrynus alces is incredibly skittish, almost aggressive, and is a very poor candidate for handling. With care a specimen can be held on a piece of wood without it jumping, but it will not sit still on a human hand. It is easily agitated and notably more defensive than any other whipspider in captivity. Startled specimens have a spectacular threat (deimatic) display involving rearing up on the hind legs and opening the pedipalps up over the body.

This is one of the only species that tries to frighten its keeper on a regular basis. The deimatic display is not entirely bluff. Although it is not among the largest of its genus and my specimens never drew blood, the spines being driven into my finger or hand on occasion were surprisingly painful. I had trouble attempting to sex the live specimens, as I usually hold animals, as seen in the various opisthosomal photographs, in order to observe the genital operculum shape. I would not suggest or attempt this with a live *Heterophrynus alces*.

Unfortunately this species did not seem to ship in very well and most of the specimens that did not die in transit only lived a few months. I was able to acquire two in March of 2009 that were probably imported in late 2008. One of them did not move on arrival and appeared dead. After a few hours it seemed to recover completely, only to really die a few months later. It had a thin, squared-off abdomen, and appeared to be a male, but did not have palps notably larger than the other which was hoped to be a female. There was no aggression observed between these two individuals the few months they were kept in the same enclosure. The larger specimen that appeared dead on arrival was in perfect shape even after it died, while the slightly smaller specimen was imported missing one whip down to the trochanter. The smaller specimen had a slightly bulging abdomen after three months and there was great hope it was a gravid female. However, it was only preparing for a molt and turned out to be a second male. Later attempts to find a mate were unsuccessful as the only other specimen located eighteen months later in the care of a hobbyist was likewise a male.

The specimen that molted lived for a total of three and a half years from the date of arrival. It shed its exoskeleton a total of three times, about twelve months apart, while the fourth was not successful. Only the first molt resulted in a visible size increase. From experiences with other species it was certainly a sexually mature male

Femoral palp spines *H. alces* (victim of 4th molt).

on arrival, but not yet full size. The left whip regenerated upon the first molt but remained undersized. (Compare the femur and tibia size in the molt photograph, which shows this specimen's second molt in captivity.) When it died it was likely the only remaining member of its species in the United States. It was kept in a humid environment with limited ventilation and fed readily on crickets and cockroaches. There were no reports of captive or wild-caught hatches of this animal.

In Brazil, *H. alces* can be found in the same forests as *H. longicornis*, sometimes even on the same tree (Weygold 1977). Immatures look similar at a glance but adults are easy to tell apart without looking at the palp spines, because *H. alces* is smaller. Also, the palpal tibia is not longer than the width of the prosoma, while it can be as much as 50% longer in large *H. longicornis*. This species' aggressive behavior carries over to nature, as it is usually solitary.

Only Guyana allows these to be exported and getting the correct, valid, and legal permit is complicated with lots of red tape, third-world politics, and hassle. Even if the paperwork is completed and hoops jumped through, the specimens have been damaged by local collectors. On top of that, the locals are afraid of them so the amblypygids are neglected at the export facility. The DOA on shipments was over 50%, with most survivors missing whips and walking legs (Gearheart pers. comm.).

Distribution: Pictured specimens are from Guyana; also Brazil and Surinam.

Heterophrynus batesii (Butler, 1873)
BATES' GARGANTUAN WHIPSPIDER

This unbelievable looking species from Peru was a "dream species" and certainly the "whipspider Holy Grail" for a long time before I finally acquired one. I chased down every whisper of a lead for fifteen years and was always too late or too unlucky. Tomberlin told me about someone in Ohio breeding this species and *Phrynus whitei* beginning ca. 1995, I read about *H. batesii* reproduction and drooled over its spectacular photographs in Schleser and Roberts (1997), and I saw a huge female that was probably this species at Insect World, Ohio, in 2001. (I thought it was a male until a few years ago because of the incredibly long palps, but in retrospect it should have been easy to see the animal in the photo is a female.) Reportedly *H. batesii* has been bred by a few enthusiasts in the USA, but I never saw livestock available for purchase prior to 2010. (That is, with the exception of the rare, unlisted offspring from the Ohio breeder I would "just miss" every two or three years. The unidentified breeder could have been in any state since I would be getting them through Arizona, but the fact they would have been coming from my own state was taunting.) Whispers suggest Bates' gargantuan whipspider has been kept in the US for a few decades, but never before has it made its way to vendor lists. The few available went to preferred or lucky customers. In the last two years, adults as well as captive-hatched young have been available a few times from a few vendors. Specimens are still an exceptionally rare find, but previously they had been unobtainable.

Although the body size is similar to large species of *Damon*, though longer and less wide, the legs are huge in relation to the body, which effectively makes it a far larger animal. This species requires medium to high humidity and a bit of space; a 10-gallon (38 L) terrarium for one adult animal or pair is a minimum. Large (not exceptional) specimens are capable of obtaining just shy of a 24″ (60 cm) whipspan (Weygoldt 2000), making it the largest documented arachnid in the world according to legspan. The captive-reared, second-stage adult male pictured has a whipspan of 18″ (46 cm), which according to the molt is not his final size, but it seems unlikely the whipspan would grow more than an inch or two at the next molt.

This is one of a few *Heterophrynus* species famous for unrealistically long pedipalps, though this specific species may well be the clear record holder for longest pedipalps among all the world's amblypygids. The outstretched palps of any full-grown male exceed eight inches (>200 mm).

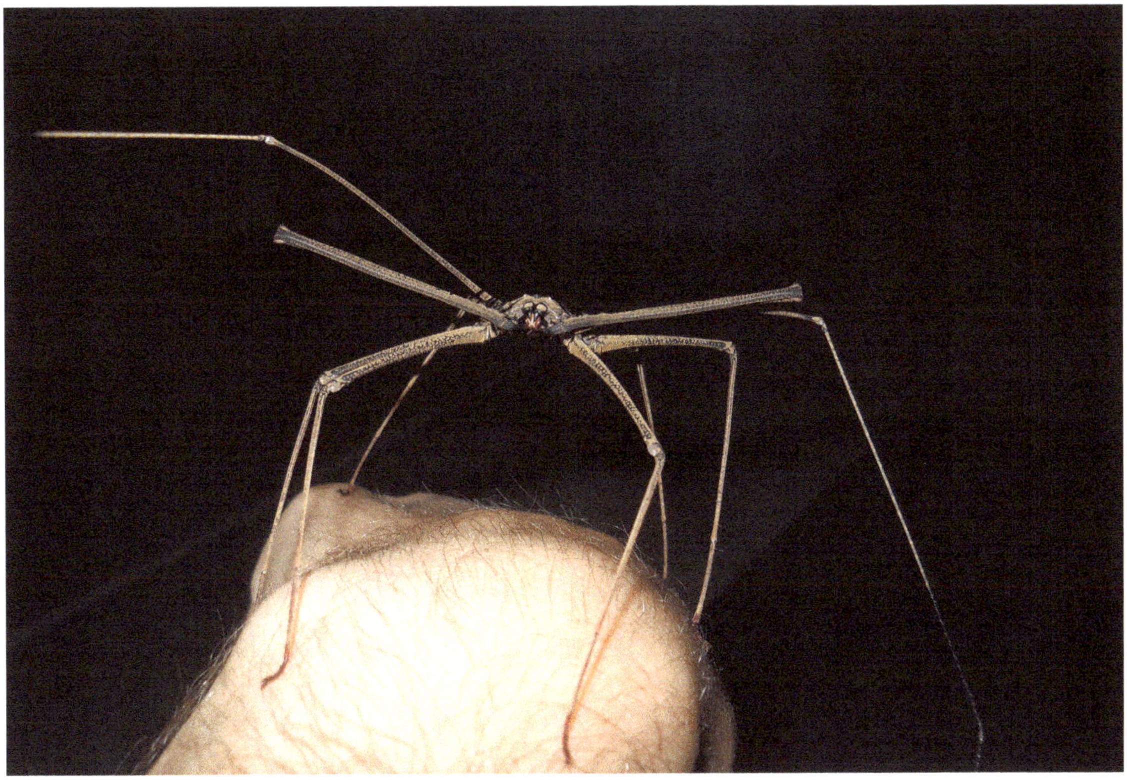

Heterophrynus batesii big smile.

Females have elongate palps but less so, and after the first postultimate molt it is easy to tell the genders apart from a distance. (They look much more similar after the first ultimate molt.) As is found in every species of *Heterophrynus,* there is a ventral apophysis on the trochanter of the pedipalp, uncomfortably close to the sharp chelicerae. The long catching spines on the dorsal surface of the palps in this genus are roughly equal in length, whereas our *Phrynus* and *Paraphrynus* have one and two short spines between the long spines.

Like its close relative discussed prior, *H. batesii* exhibits three-dimensional movement of the pedipalps as they are reared back up over the prosoma. Individuals with long palps sometimes hold them bent up at the elbow while forward and open for prey capture. The secondary posture of the pedipalps is seen when prey is placed in the cage. Hunting specimens hold the pedipalps back up over the body. This is normal behavior, but they are very skittish and capturing the hunting posture on film is difficult. The slightest disturbance causes a change of posture. The descriptive photo included here took many tries and is not as descriptive as hoped. I have a hundred other photos with flash glare or foggy glass, or they are simply out of focus. Immatures often hold the palps forward and slightly bent at the elbows when at rest. This is similar to *P. mexicanus* and likewise palps are retracted at the slightest disturbance.

Despite the larger palps and bigger body, it does not have a notable threat display in captivity like *Heterophrynus alces.* It is more docile for handling, but specimens spook easily and can still jump, even

Original F1 L2 *Heterophrynus batesii.*

F1 molt to maturity at 13 months *H. batesii.*

if held on a piece of wood. Risk of death from falling is uncertain, but other species close to a molt that jump three or more feet (>1m) to a hard floor may be found dead the next day despite no visible injury. Fat females of any species might crack or damage the abdomen at half that distance. I have twice observed the adult male run off the wood, up the back of my arm and onto my back faster than it is possible to react. In this case the animal stays on my back and I have to get down on my hands and knees, which convinces it to look for refuge and finally comes within reach. Talk about a pet that keeps you humble . . .

Whipspiders are often defined by their flattened body. However, not only developing immatures, but also adult females of *H. batesii* can have fat opisthosomas that are as tall as wide. The body color of this species is difficult to photograph because it changes with lighting. In dim light the animal is dark brown with reddish tibia, under bright room lighting it looks suede, but under a strong flashlight a sort of golden fluorescence like a reflective road sign is cast by the reddish, powdery coating. The legs are also covered in this coating, along with glistening, golden setae.

Specimens can be kept in a 10-gallon (38 L) or larger aquarium with a glass lid. Large pieces of cork bark should be set in the cage at a slight angle for molting. Crickets are a favorite food, but chew on molting amblypygids. Even minor damage that occurs within a day or two of a molt will nearly always kill a whipspider. After nine months I lost two on the same day to a few crickets hiding in a small cavern under the edge of a piece of cork bark. A few 1/3″ (8 mm) crickets were hiding unseen. For many months afterwards, only *Schultesia lampyridiformis* were used as prey.

This species is said to be more aggressive towards conspecifics than some other *Heterophrynus*. Many reports as far back as Beck (1968) suggest *H. batesii* adults are usually found alone or less often as pairs on a tree, but a tree with *H. longicornis* can host a male, female, and a number of immatures (Weygoldt 1977). This behavior would seem to correlate with the much shorter palps of *H. longicornis*, since long palps are for sparring. When I finally located

Color with flash, *Heterophrynus batesii.*

Color without flash, *Heterophrynus batesii.*

and purchased a wild-caught "pair" they eventually turned out to be an old adult male and a young adult male. There was no aggression noted over a sixty-day period of cohabitation. The large, wild, adult male died at this time. It had no physical damage and seemed to have passed from old age. A month later the fat "female" I hoped might be gravid molted into an obvious male. This story may seem like a repeat of the *H. alces* experience, but these were bought as a pair and fortunately, this time, the story did not end here.

When I acquired the "pair" (late August 2010) I also purchased some captive-hatched immatures that had thin opisthosomas, indicating they had fed little, if at all, after leaving the mother, at most weeks earlier. The early instars of this species are very colorful. Like other whipspiders they start out with red pedipalps, but they also have a yellow margin circling the prosoma and yellow markings at the leading edge of the dorsal segments of the opisthosoma. The brown base color darkens after a few more molts and the yellow and red markings fade away as they transform to dark suede. The yellow margin and red palps become greatly reduced after reaching 5th instar and are not discernable at 6th instar.

Of nine immatures, one died early on, and another two had been eaten by small crickets minutes after the molt to 5th instar, possibly within minutes of each other. After a little over a year, the largest had gone through five subsequent molts (equivalent to 7th instar, including the molt off the mother, but not including the development from egg to postembryo inside the broodsac). I was surprised by the tiny males, as they began forming spermatophores the last few days of December 2011. Apparently these were 1st stage adults, not 7th instar immatures. Before this time sexing seemed impossible as the male's pedipalps were not yet long and the genital operculums looked similar. I also had no baseline for comparison, as I had not yet seen the genital operculum of an adult female *Heterophrynus* species.

At about this time the two smallest were eaten (unrelated to molting) as the more developed specimens were much larger and had become sexually mature. Unfortunately one of these was from a later, unrelated brood that had been added to the group three months earlier because it had nearly caught up in size.

Reproduction: I was very worried about handling before it was obvious they had hit initial sexual maturity. They seemed fragile and might have dropped legs or died from handling. Healthy specimens are skittish, run quickly, and jump. I had tried to check the operculums a number of times by looking up from the side, but without recognizing a difference in younger immatures. They are much more difficult to pin down than *Damon* or *Phrynus*, and I was more concerned about hurting them. My experiences trying to handle the violent and skittish *H. alces* were also a consideration at the time. With only four left, the risk of killing them from aggressive handling was no longer outweighed by the safety of knowing which was which. After gaining familiarity with the operculum forms and pedipalps (female palpal tibia curve outwards at a much greater angle towards the hand), it may be possible to sex immatures as early as 5th instar. The four largest remained—two male and two female.

It is normal for amblypygids to reach initial sexual maturity at a small size, but they had matured a little smaller than I thought they would. I was also surprised because Schleser and Roberts (1997) said the time to reach maturity could be two years or longer (not "or much less"). I observed a few spermatophores made by the small males at barely sixteen months of age, the last few days of 2011 and first week of 2012. This meant the larger specimens had begun maturing with the most recent molt a few months earlier at thirteen months of age. Even after seeing spermatophores, I did not anticipate successful reproduction for a year or two due to size, time, and conditioning requirements. First-stage mature female *Damon* rarely produce eggs after mating, and when the rare one does, the brood usually fails and the female must molt again before mating.

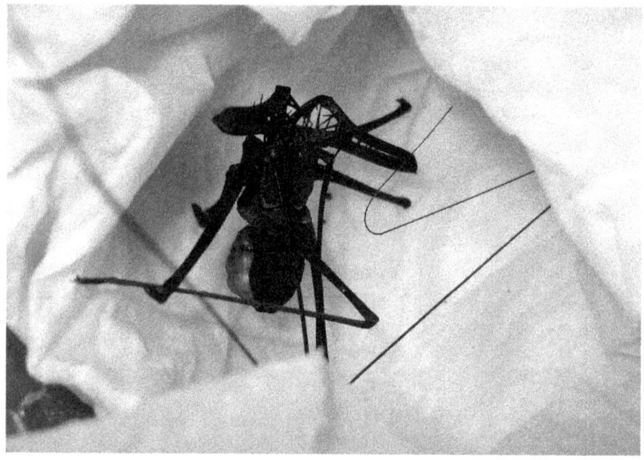

Note autotomized legs and the broodsac is still not chocolate brown at thirty days post extrusion. *Heterophrynus batesii* female 1.

One fat female had been moved to her own cage and kept at 80° F (27° C) for a few weeks. She extruded a broodsac only eighteen months (February 19, 2012) after she herself hatched, which was amazingly rapid. The following day, five of her six walking legs were missing from the tibia down. There were no other animals in the 10-gallon (38 L) terrarium beyond some barely visible stone centipede babies and springtails. There was no metal screen anywhere and I can only guess the high temperature may have been the culprit, but otherwise the cause of the problem was unimaginable. Malfunctioning autotomy is seen in the house centipede *Scutigera coleoptrata* and certain phasmids (when the animal is stressed or dying the legs start falling off). She was moved to a 32 oz (1 L) deli-cup with an inch of moist dirt covered in paper toweling, and limited ventilation also kept at 80° F (27° C). I was hopeful because she seemed functional beyond the missing legs. She seemed to decline no further in health for a little over a month. Even after more than a month the broodsac membrane never fully sclerotized. (The broodsac membrane of a healthy female turns from clear to a dark caramel brown in less than a week.) She died on day thirty-five and took her developing eggs with her. Eggs and first instars cannot survive without a healthy mother. Amblypygids can be terribly trying at times.

Broodsac only begins to darken on day five, *Heterophrynus batesii.*

This terrible failure seemed to offer no hope for success and not a scrap of valuable information. I did not realize until much later as I compared data and photos that the broodsac membrane had never fully sclerotized. This odd failure seems to offer valuable insight into the chemical processes within the broodsac. It is interesting because it suggests whatever her health problem was, the broodsac remained chemically connected to her. The fluid connection is known, but this showed the chemicals in her body are also responsible for the broodsac membrane sclerotization that occurs about five days following formation. If she was still not chemically connected and the sclerotization process were a result of chemicals already present in the membrane, on extrusion it would have displayed a normal coloration as it "dried," somewhat like a mantis, cockroach, or water scavenger beetle ootheca whose external surface dries a certain color no matter how badly formed.

A few hours later the entire membrane is caramel brown.

The remaining female appeared ready to extrude eggs when the first female died, so there was great fear she would fail in a different or similarly spectacularly strange manner. However, I placed her in an area where temperatures were around 72° F (22° C) in response to the other female's failure and past experiences with other species. She extruded eggs on March 20, 2012, which hatched on August 29 and at no point did any of her legs fall off. The dispersal molts (from 1st to 2nd instar) of her offspring took place from her body on the evening of September 9 to the morning of September 10.

Heterophrynus batesii F2 L3 coloration.

According to Schlesher and Roberts (1997), gestation takes only three months, which sounded very rapid considering *Damon diadema* rarely take less than four. The gestation time I recorded was five and half months. I realize now they used the word "approximately" and likely were relaying secondhand information or information for other Amblypygi, and their gestation temperature may have been a statement of average surface temperature in the region rather than laboratory gestational temperature. Schleser and Roberts implied they observed *H. batesii* development in the laboratory, but a more critical look at the article reveals they not only put qualifiers on stated times, but also offered no specific data and bounced between the subject species, multiple species, and the entire order in their references, usually without explaining when they were switching. The reproductive description was ". . . based upon observations in nature and under controlled conditions." But which parts of the description? They also showed and discussed failed broodsacs without describing the barest of parameters and did not state which pictures were natural or if their wild-caught females were captured with the broodsacs already on them.

Observations during attempts at capturing very young 2nd instar immatures show great care in handling must be taken at this time. Defensive autotomy at this instar is highly developed. I have never pulled a whip off an animal with my hands, but when I was trying to catch some two-week-old 2nd instar *H. batesii*, two of them each threw a single whip at me. I did not even have a chance to grab them yet. When

Jan 7, 2012, Spermatophore from one of the small males.

almost as well as crickets. (*N. cinerea* still hide well and those not eaten within a few hours may never be eaten or consumed weeks later.) A calcium-vitamin dusting regimen was also begun shortly after, since this prey species is easy to dust. *Dorylaea orini* and a few *Eurycotis* spp. altogether made up 30-40% of the diet this generation (and only a couple were offered to the parents). I expect this generation to mature more rapidly, and hopefully larger, than their parents, due to improved prey quantity and quality. (The largest are presently on track to mature at ten months versus the

they are very young they are also very likely to jump. If a piece of wood with individuals on it is held up, they may stay on it or jump right off, so it should always be held directly above the open cage.

Molts often take place at night, but a number of molts have been observed during the day. The day night cycle provided is 14 hours daylight and 8 hours dark. However, lighting during the day is indirect and dim.

Of the 25 immatures to emerge from the first broodsac, one died a few days after dispersal for no apparent reason. It was measured and the 2nd instar whipspan was already 4.25″ (108 mm). The survivors fed well on small crickets, *Dorylaea orini*, and *Eurycotis decipiens*, and grew rapidly, if somewhat staggered. No others died and by January 23, 2013, (135 days post-dispersal) the first began molting to 5th instar. Habitat temperature was approximately 75° F (24° C). The parents from 2nd instar to adult had never been fed *Nauphoeta cinerea* because it did not prove to be a very useful prey with *Damon* or *Phrynus*. I only discovered this generation, at 3rd instar, through a feeding experiment, will hunt *N. cinerea*

2nd, 3rd, 4th instar, development was growing more and more staggered prior to improved feeding (afterwards the very first and last were less than a molt apart).

parents' thirteen.) The stragglers were able to catch up with improved feeding and even the smallest was 4th instar before the end of January. Stragglers normally grow further and further behind. Amazingly, every specimen was the same instar (5th) on April 6. All had finished the molt to 6th instar by early June 2013. The first to molt to 7th instar was on June 28, 2013. I just saw this fresh molt; it is a little over 9.5 months.

The mother molted on October 31, 2012, sixty-three days after the hatching of the broodsac (eleven days shorter by dispersal). She was reintroduced to a male on December 7 and mated that evening. The second-stage female began extruding her broodsac February 19, 2013, at 2 p.m. The cage was directly backlit with the lights off at 10:30 p.m. My efforts at photographically documenting the extrusion process may have doomed them, as the female was spooked and dirt became imbedded in the broodsac membrane. I messed up a few *Damon* broodsacs in the past trying to get photos, but this was on a whole new level due to my great love for this otherwise unobtainable species, severely reduced chances of establishing a viable stock, and theoretical loss of value. (Even if the photographs had been spectacular, there could be forty or more offspring individually more valuable to me than a good photo.) To add insult to injury, multiple pictures were all turning out foggy and I did not discover till days later that the retracted camera lens had become dirty.

Initial development of the broodsac has been normal and there appear to be no signs of punctures. I closely observed broodsac sclerotization and noted it remained clear for the first four days while the excess fluids were drawn in by the female genitalia. She maintained an inverted position day two to five. On the afternoon of day five the broodsac began to turn yellow and within a few hours of the initial change it developed a deep caramel color. Day 60 postembryos noted and Day 120 the broodsac appears healthy. Hopefully the attached substrate will not cause eventual failure.

Longevity reports for this species are presently extremely limited. The small male that molted late 2010 and grew in size greatly, molted again in October 2011 with a notable but far less impressive size increase (maybe 30%). It molted again in October 2012 with no size increase. After this molt it seemed weak, the legs were a little crooked, and it died a month later in November. This male may have molted to maturity in nature a year earlier, which altogether would be less than three and a half years for a mature specimen. I only know of a single original imported adult still alive as of January 2013, which is also a male. Longevity data for imported specimens of *H. alces* and *H. batesii* maintained in captivity by various hobbyists seems to contrast greatly with the rare reports of advanced longevity, such as one female specimen of an unidentified *Heterophrynus* sp. that lived for 11 years as an adult (Seiter pers. comm.) or the less specific reference to ten years for *Heterophrynus* in Weygoldt (2000).

Female *Heterophrynus* from Peru at Insect World 2001.

In Nature and Imports: This species is found on huge tree trunks in the rainforest. In the area of Peru where original stock was collected the locals refer to this species as "congrejo," which means crab. They were collected at night near the base of the very large *Ceiba pentandra* trees, known locally as Lupuna trees and elsewhere as Kapok trees. The *Heterophrynus* would be sighted on the tree trunks about three to five feet (1-1.5 m) off the ground, facing downward with whips outstretched presumably to detect food or danger. They were always within the massive buttresses and could retreat quickly into fissures upon approach. Large quantities were never collected and they did not seem to live in colonies (Bollinger, Floyd and Martinez pers. comm.). Schleser and Roberts (1997) also report that, with exception of pairs, the adults appear to be solitary and are found no higher than 8 ft (2.4 m) from the ground.

2010 imports came from an area known locally as Las Palmas in the state of Lareto in northeastern Peru. The initial survival and health of the animals was good, and appendage loss minimal. A number of pairs were set up for mating with about a third ending in the loss of one animal. (This may have been partly due to improper sexing as at least one "pair" were different aged males.) A few broods of healthy captive-hatched young were produced from the wild pairs mated in captivity. Unfortunately, the exporter in Peru lost his export license later in 2010 so new exports are unlikely (Floyd pers. comm.). If present stock is not maintained and bred, this species will shortly again be as available as the real Holy Grail.

Broods: 25-50 young are produced about once a year. Gestation of the captive F2 2012 brood took five and a half months. 1st instars molted eleven days after hatching over a 16hr period.

Distribution: Heterophrynus batesii is known from Brazil, Columbia, Ecuador, and Peru.

Heterophrynus elaphus Pocock, 1903
PERUVIAN GARGANTUAN WHIPSPIDER

The name *elaphus* means deer and certainly is meant to compare the spiny palps to the antlers of a male deer, but both genders of *H. elaphus* have similar looking palps. More specimens of this whipspider have been imported into North America than any species other than *Damon diadema*. This was the first *Heterophrynus* I acquired. Unfortunately specimens are always killed and dried prior to leaving their country for the dead stock trade, (The coloration of the dead specimen pictured is likely quite different from a live specimen,) The short palps and appendages are reminiscent of *H. alces*, but this species is half as much larger. Whipspans on two specimens I checked were approximately 18″ (46 cm). Both may have been females, but only parts of the genital operculums were exposed since they were mounted on cork, the abdomens deflated, and there is no notable sexual dimorphism of the palps in this species. The legspan of the walking legs of the pictured specimen is 9.25″ (24 cm). The legs are a little longer and thicker than captive-reared *H. batesii*, though the body is similar in size. I have not measured additional specimens but an especially large specimen on display at Harrell House of Natural Oddities in Santa Fe, NM, is documented at very close to 24″ (610 mm) and a body length of 41 mm, so this species certainly is able to compete with *H. batesii* and *H. longicornis* for the largest title.

I spoke to a dried stock importer who has imported these over the years and she told me how she visited the exporter in Iquitos, northern Peru, and experienced seeing a surprising behavior of these gargantuan animals up close (a few decades

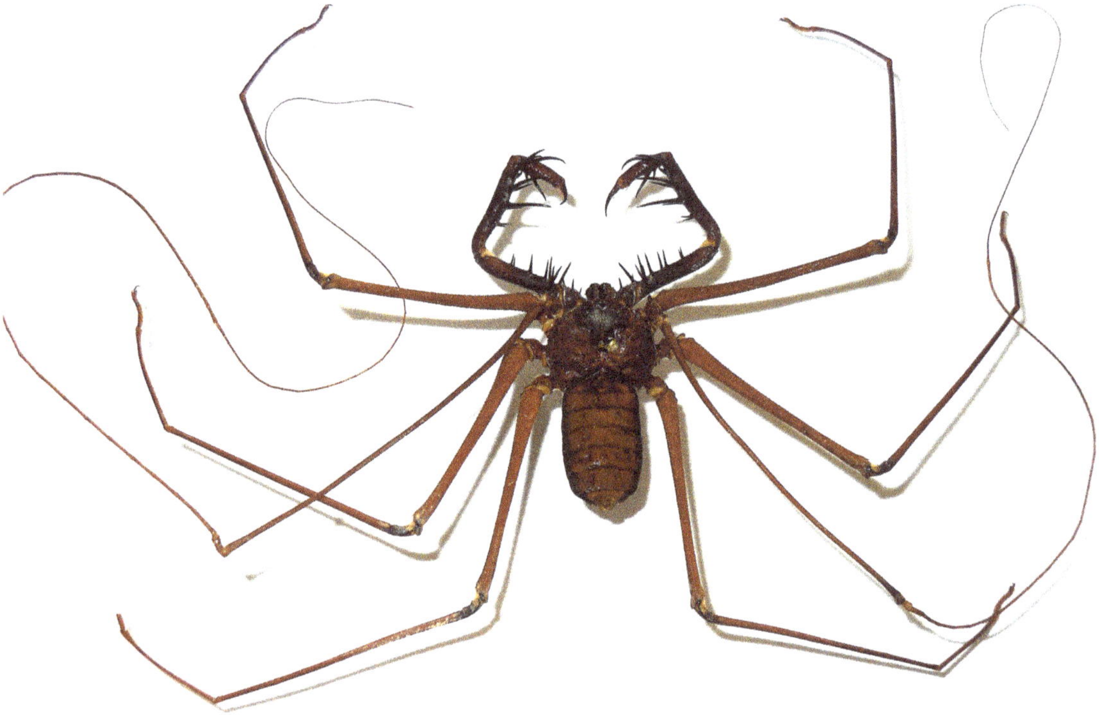

Heterophrynus elaphus dried specimen.

back). It was night and the guide took her tour group through the forest to a very large tree with a rotten hole in its side. The hole was the size of a watermelon, indicative of the size and age of the tree. The guide stuck a long stick in the hole and rapped on the sides of the cavity. Slowly, whips first, a dozen massive *H. elaphus* crept out of the tree cavity one after another. Their movement was so methodical and measured that it came as a great surprise when a few jumped onto nearby trees and onto two members of the tour group trying to get a close look (Hurt pers. comm.). Most amblypygid jumps are more of a release and drop than a loaded jump like a springtail or grasshopper. I have not yet observed jumping in *Heterophrynus* adults, but my goal has been not to spook them. However, I have seen some *H. batesii* large immatures jump from one piece of cork bark to another or from my hand to the decorations. They lower their body and then push up rapidly with all legs to end up on the opposite surface without losing notable height. This miniature representation makes it easy to imagine just what the tour group observed on a grand scale.

Distribution: Peru.

Paraphrynus mexicanus (Bilimek, 1867)
Sonoran Desert Whipspider

Our native *Paraphrynus* species are easy to differentiate from *Phrynus* due to the spine structure of the pedipalp tibia. There are two short spines between the two longest for *Paraphrynus* (see fig. D), but just one short spine for *Phrynus* (see fig. E). Specimens have a consistent body coloration—jet black radiating out from the body halfway through the femora, with red brown continuing to the tarsi. Unlike *P. marginemaculatus* that cling tightly as they run all over a piece of wood and rarely jump, these are three-dimensional and hold the body up on the long legs and rarely lay completely flat even at rest. Specimens in enclosures spend the day on the side of bark pieces a whip length or more above the substrate, unlike *P. marginemaculatus* that hide in narrow cracks. They readily jump, or drop off, pieces of wood being held if spooked. With a body length just under an inch (23 mm), they are not terribly larger in body size than the more familiar *P. marginemaculatus*, but the whipspan can be much larger. The whipspan of my largest *P. mexicanus* is 8″ (203 mm) versus the largest *P. marginemaculatus* at 4.5″ (114 mm).

Closeup, *Paraphrynus mexicanus*.

Phrynus marginemaculatus and *Paraphrynus mexicanus*. These are both large adults, note the *P. mexicanus* body is not terribly larger but the legs are comparatively gigantic.

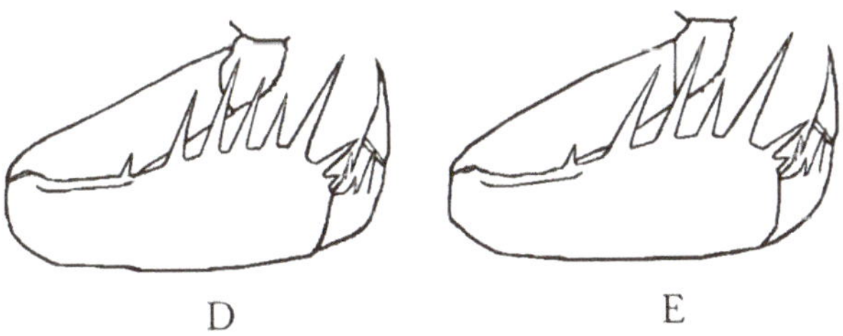

Phrynus vs. *Paraphrynus* pedipalp tibia diagram.

This species had been nearly unheard of in collections, though it is a common species that is probably found in huge numbers throughout Arizona. Specimens have shown up in captivity over the years, but they have only made it to vendor's lists in 2012. In the wild, Sonoran Desert whipspiders have been found on the ceiling of round-tailed ground-squirrel burrows (Werner and Olson 1994), inside water meter boxes, and trapped after wandering into garages. They may be found in cellars and houses under sinks, but mostly in newly-developed areas because their habitat has been replaced with human dwellings and they have nowhere to go. Specimens are most commonly reported from less brutal areas and near water sources, but also occur in very hot dry areas including Phoenix (Stoops and Martin 1995). The Tucson area is often brutally hot, but *P. mexicanus* are found in protected areas among the foothills of the Catalina and Rincon Mountains and mesquite forests in

The 3-dimensional *Paraphrynus mexicanus*.

Paraphrynus mexicanus will lay somewhat flat.

Tanque Verde Valley (Kirby 1986). Six of the animals depicted here came from Tucson and five from the Arizona side of Lake Havasu on the Colorado River. The Lake Havasu specimens were collected together from a small shed. Since this species is not found in California, the Colorado River would appear to be the westernmost border of the range.

They often face each other with the palps extended, but the movements do not appear rapid or aggressive, and they do not usually touch palps. Usually the palps just hang open and do not move for hours on end. They may face towards each other or in any direction. Palp extension behavior appears to be the same between individuals of the same or different sexes, as well as with juveniles less than half the size of adults. A different, uncommon behavior involving the palps has been observed. Twice, a pair was seen in behavior that seems like early stages of courtship but there has been no spermatophore observed, so it could possibly be aggression. Both times it was a male and female and none of the large specimens have exhibited any damage so far. The pair extends palps towards each other, the male waits five to thirty seconds and then grabs the female palps—she then recoils and backs up. The movements are quick but rather slow compared to prey capture or standard aggression, and the process is repeated over and over again, possibly for hours. This behavior was only seen during the first six weeks, which is suggestive of courtship observed in other species.

I kept specimens together for ten weeks and was impressed, not by the lack of aggressive behavior despite some specimens being very tiny, but because they would almost never be found alone. They often switch hides and positions in the cage, while groups contain constantly shifting gender ratios and sizes, but specimens are almost never found by themselves. Additionally, they cluster when threatened, like early instar *D. diadema*. However, I had reduced

Hunting posture of *Paraphrynus mexicanus*.

food somewhat to try to get them to eat the remaining *Deropeltis* cockroaches and after ten weeks I found one small specimen absent from the habitat and a lonely whip on the substrate. (A far smaller specimen which had already been missing a whip when received, was still present.) If this animal was not molting this would be the only time I have personally observed indications of cannibalism unrelated to molting beyond early mature *H. batesii* feeding on developmental stragglers. Adequate prey was added and no additional specimens have disappeared in more than seven months.

The enclosure, a 5-gallon (19 L) terrarium, was kept on a bottom shelf from mid-November 2012 through March 25, 2013, which is around 69° F (21° C) during the winter and early spring. They were kept cool for conditioning, since they arrived

There is always some variety in natural specimens, the right palp on this irregular specimen has an extra spine between the large two, *Paraphrynus mexicanus*.

Paraphrynus raptator immature specimen with red palps. © Sam Floyd

Paraphrynus raptator. © Sam Floyd

Paraphrynus raptator from southern Florida. © Sam Floyd

rather thin and too warm of temperatures sometimes prevent animals from gaining weight. They fed readily and fattened up nicely over the first few weeks. The enclosure was moved up two shelves in the early evening of March 25. This shelf is a few degrees warmer—about 72° F (22° C) at this time of year. The very next afternoon I witnessed late-stage mating and spermatophore production. An hour later, all signs of the spermatophore had been gnawed from the bark, so it might have taken place before. However, all four of the adult females displayed visible increases in feeding within days (primarily of small *N. cinerea* that had long been hiding in the substrate, but also small crickets) and a partial mating event was seen five days later. All were stocky and healthy, but these females were the only specimens to develop notably swollen opisthosomas over the next week. In this species the sternites are variably see-through and in just two weeks the ova had grown notably in size and turned green. A fifth, very small female (16 mm) also has eggs growing ready for broodsac formation.

Distribution: Sonoran Desert area east of the Colorado River. Arizona (USA), at least the southern half of the state, including La Paz, Maricopa, Mohave, Pima, and Pinal counties; also Mexico.

Paraphrynus raptator (Pocock 1902)
Archipelagic Whipspider

This is the only other US member of the *Paraphrynus* genus. It is supposedly only found on Key West, FL, where natural habitat is in short supply, but the specimens pictured here were found in an undisclosed location on an island about 60 miles north of Key West. Unlike *P. mexicanus* it is probably rare and certainly has a smaller range. I have been told this is our largest *Paraphrynus* species in body length (up to 1" or 25 mm), but extreme rarity makes this difficult to confirm or deny. It certainly appears to have much shorter legs and whips than *P. mexicanus*, even were the body a few millimeters longer. It appears to have shorter legs than our native *Phrynus*

Coral rocks with easy escape routes for *Paraphrynus raptator.* © Floyd & Bollinger

Phrynichus exopthalmus female, compare palps and chelicerae to male.

Phrynichus exopthalmus male.

operculatus. Some of the *Paraphrynus* species can be easy to tell from *Phrynus* at a distance, as a result of the long legs, but there are numerous *Phrynus* species and some have very long legs, which is why the palp spines are a useful identifier. I am only aware of one immature *P. raptator* specimen that ever made it to a vendor list in 2012. A captive-breeding attempt with wild-caught adults ended in cannibalism (Bollinger pers. comm.).

In the Wild (Bollinger and Floyd pers. comm.): The habitat where *P. raptator* were found is a limestone coral rock with small cavernous holes, where they rarely ever come near enough the surface to be collected, somewhat paralleling the rocky habitats of *P. mexicanus* in the Sonoran Desert. The habitat is dominated by both black and red mangrove trees *Avicennia germinans*, *Rhizophora mangle*, gumbo limbo trees *Bursera simaruba*, as well as Brazilian pepper *Schinus terebinthifolius*, 100 yards or so out of reach of the highest tides. The habitat and behavior of *P. raptator* differ from the other native Florida whipspider. While both display the same thigmotactic response of running under an object of cover as it is being flipped, *P. marginemaculatus* typically cling underneath, while the *P. raptator* has a tendency to drop off and run towards a deeper hole. The *Paraphrynus* live in smaller communities; there are never two under a single rock, whereas a number of *P. marginemaculatus* can be found under a single log.

Distribution: Cuba, Honduras, Mexico, Florida (USA) (Harvey 2003).

Phrynichus exopthalmus Whittick 1940

The species name probably refers to the widely spaced median eye placement on the prosoma. The body is covered in a thick brown powdery material and this creature is certainly as resistant to desiccation as any *Damon*. Other than the tan body color, the main feature that sets these apart from *D. diadema* at a distance is the lack of spines on the extremely long and thin palps. There are some tiny spinelets but even the large spine normally located near the base of the tibial hand in other *Phrynichus* is missing. This spine structure is sparse even by *Phrynichus* standards, which would seem to place it among the *Euphrynichus*, but Weygoldt (2003) documented the spermatophore structure and courtship, which along with the female genitalia place it in the *Phrynichus ceylonicus* group. Male and female maintain close contact during the mating dance and the male presses the female onto the spermatophore. This is unusual, and likely the reason it has been able to mate in crowded captive conditions. I have only seen this species twice in US collections, misidentified as *Damon variegatus*, as they were brought in from Tanzania under that name, probably in 2009-2010. (It would seem that "Damon variegatus" is employed as the common name for all African Amblypygi.) This is a very handsome, larger species, maybe 75% the mass of *D. diadema*.

Males show very little aggression towards each other in a somewhat crowded communal habitat, despite close proximity and impressively oversized palps commonly associated with sparring. In older females the palps are very long and narrow and the elbows extend further than the femur of the front walking leg. In older males, however, the folded palps are much more impressive and can grow to twice the length of the front walking leg femur. A mature male's chelicerae are notably thicker and more massive than those of any female. His chelicerae are so greatly enlarged that the prosoma curves up behind them. Oversized, slicing chelicerae would seem to be an adaptation for killing other males, but may relate to courtship. It is not an adaptation for catching prey since it is only found on adult males.

The group of specimens shown (among those originally labeled *D. variegatus*) is part of a small, crowded breeding colony that has been maintained for greater than three years in captivity and is presently composed entirely of captive-hatched individuals. Only about one adult is lost a year, due to aggression related to molting cannibalism, and generally only when prey items have not been offered recently. The female's broodsac appears similar to *D. diadema* in that her pleural folds only cover the outer edge, so the eggs are visible through the semi-opaque broodsac membrane. Multiple groups of immatures have been reared together safely in groups kept separate from the parent group (Hodge pers. comm.).

Broods: 20-30.

Distribution: Africa; Cameroon, Congo, Kenya, Rwanda, and Tanzania.

Phrynus barbadensis molting from a screen lid. © Matt Kaplinsky

Phrynus barbadensis (Pocock 1894)

BARBADIAN WHIPSPIDER

In theory this species does not get much bigger than *P. marginemaculatus*, but in practice it is a huge animal that dwarfs our

Phrynus barbadensis pair, male at front right.

miniscule Florida native. The bodies of three adult female specimens I have seen were 0.87-0.91" (22-23 mm), but they have a huge, wide prosoma that an entire *P. marginemaculatus* could fold up inside. The whipspan is much more expansive at 8" (203 mm). While similar in measurement to our *P. mexicanus*, it is far stockier and more massive. The pedipalps are thick and fat like boxing gloves, while the legs are thin, short, and dwarfed by the body. The body of mature females can be nearly solid black and smooth, resulting in an animal that looks like no other in culture. Specimens do resemble an extremely large *P. marginemaculatus* or late instar *D. diadema*, but the massive, squat prosoma gives it a unique appearance. The front of the prosoma has a raised lobe behind each chelicera, which is only notable on the adult male and is barely pronounced on large immatures or females. The chelicerae are held abnormally far apart at rest, so the animal can press tightly against the bark without palpal tarsi getting in the way. There is almost space to fit another chelicera in the middle because the palpal distitarsi rest here. Some people call this the "camel toe whipspider," but there are no camels in Barbados.

Barbadian whipspiders seem very well adapted to a hidden life on the side of a tree. During the day they do not move and are the only whipspiders I have seen that are sometimes found resting tightly against the upper surface of wood pieces in the habitat. I have one female with lighter brown markings on her opisthosoma that is often found tightly adhered to the upper surface of the bark, while the others are almost never on top. Specimens stay flat and motionless, but can move like lightening if the legs are pushed on enough times to annoy them. The log or bark an animal lives on can be pulled out of the cage and waved around and the animal is not going anywhere. (Do not take this literally, as even the most surface-hugging amblypygid could still possibly jump.) Specimens even keep a low profile during hunting and feeding, which usually only occur at night. Most species lift the chelicerae away from the surface as they feed, but these keep their faces low to the ground with only the palps angled upwards as a sign of feeding. Two-dimensional behavior carries over to mating and the incredibly low angle of the spermatophore.

Fortunately in this species the shape and indentations on the genital operculum are very different between the two sexes (like *Damon*), so it was easy to be certain the initial two were a pair, even though they were still small. *Phrynus marginemaculatus* genders are far more difficult to identify this way and can even be difficult to differentiate in old adults. Like most species, *P. barbadensis* put up with being grasped between the finger and thumb, and stay still while being checked without biting or stabbing. (However, this assumes one is not putting the fingers directly on the chelicerae or allowing the palps ready access.) They are easy to pin down since running away and dropping off surfaces are not part of their normal defense.

Specimens can be maintained with high humidity and limited ventilation. They are good feeders and do well on crickets and various cockroaches, even *Nauphoeta cinerea*. Adequate molting areas include screen lids and angled cork bark or similar rough surface at approximately a 45° angle and four or more inches (102 mm) in total height.

This seemed to be a very promising species as many were imported and the initial survival was excellent. Babies hatched from wild-caught females were inexpensive and seen for sale. I acquired a pair from one of the initial imports in July 2009, but they were immature and the female had a deformed body, missing whips, and part of a leg. Despite great handicaps she fed and kept her body oriented correctly most of the time, but did not look like she could hang on to properly exit her old exoskeleton during a molt. On September 7, six weeks after arrival, I was amazed when she molted and grew back her missing body parts (McMonigle 2012b). It was surprising because she did not look heavy enough to be ready, nor healthy enough to succeed. However, I was unfortunately also surprised by what followed immediately after. Just after the successful molt and whip regeneration, while she was still white in color, the male ate half her body. This certainly took place just minutes after her molt because sclerotization of the exoskeleton is an active chemical process arrested at the time of death. There was only a single pair in a 10-gallon (38 L) terrarium with numerous hides, but apparently this is a species that should never be kept together during a molt. (Keep in mind *D. diadema* and *Heterophrynus* pairs molt right next to each other without issue, so this problem was not considered and may or may not be status quo for this species.) I acquired a very old pair in late 2010 from the initial

Size comparison between *Phrynus marginemaculatus* and *P. barbadensis* adult females.

imports but they were kept together since neither looked like a molt would be forthcoming—they were kept together by the previous keeper for more than a year. The female was rickety in movement and died of old age within a month—still in good shape for a dried mount. The male continued on till summer 2012. In January of 2013 I acquired two pairs from a 2012 import and each specimen was given its own cage. Each ate two or three prey items the first night. A few days later I learned about a paper by Alexander (1962) which says *P. barbadensis* (referred to as *Admetus barbadensis*) only exhibit threatening behavior towards each other and such behavior is independent of gender. Hopefully I can discover if aggression translates to early development brood cohabitation in the not too distant future.

Mating Experiences: After conditioning for a month I decided to place the mature male in with the heavier female (female 1) one afternoon. They started out with whip taps and whip vibrations and lifted the palps in a low V. After fifteen minutes they were grabbing each other, which appeared aggressive, and the female displayed some hand-wiping behavior. After ninety minutes they had settled, but both were intermittently flicking the whips, so he was moved back to his enclosure. I introduced him to the thinner female (female 2) the following day and there was whip flicking for about fifteen minutes, the palps were not opened, and both animals quickly

Note the amazingly intricate structure of the spermatophore. *P. barbadensis*, April 29, 2013. (Top photo is base triangular cross section.)

settled. Over the first two days they were usually found on opposite sides of the bark, but were seen facing each other a few times. I only was able to check in on them a few times a day, but found a spermatophore later on the second day, so mating may have occurred as early as the first night. On days three to seven they were found together, though seldom facing, with legs nearly touching the entire time. One time the male was even seen resting with her against the upper surface of the bark where the female often spent part of the day. On day seven the male was moved back to his enclosure only because it was empty and waiting for him.

Three weeks later I introduced the male to female 1 again. It was noon. This time I observed similar behavior but gave it a few hours to see how it played out. I wondered if the behavior might be more ritualistic than angry, and the female did not seem quite as agitated this time. The male was seen to do nearly all of the whip flicking, while the female again would usually open her palps first and the two would grab at each other, but never try to hold on (like *P. mexicanus*). After approximately three hours (~3 p.m.) the male formed a spermatophore on the bark and the female promptly tore off the head of it using her genital operculum. The high level of interaction seemed to have been because female 1 was more receptive rather than less. However, twenty-five hours later they were found fighting (vigorous palp attacks without whip flicks) and the male was placed back in his cage. The tarsus of one of his whips had become stuck, bent back behind the tibia, but it was unhooked, retained function, and was not later autotomized.

Five weeks later the male was reintroduced to female 2. She had only eaten one large cockroach a few days after mating previously and showed no signs of increased feeding or weight gain that is common after successful fertilization. She completely ignored his introductory whip flicks, but whether she remembered him or had no interest would be hard to prove. However, they shortly resumed close contact as seen days 3-7 of the first cohabitation. Additional spermatophores were seen over the next three weeks and large eggs became visible through the transparent abdominal sternites. Broodsacs formed June 11 (female 2) and June 27 (female 1).

Eyes and palps of a female *P. barbadensis*.

Captive and Wild Availability: Females holding broodsacs in captivity have not been rare but with this species the eggs often disappear after a month or two. One female produced approximately 30-40 eggs per broodsac, two times, that appeared to develop, only to be eaten prior to hatching. A partially eaten eggsac was taken from the mother as she was eating it, and there were exposed, fully-formed postembryos. They were placed on a moist paper towel, but did not become active (Kaplinsky pers. comm.). Even if they had survived, they would never have lived through the first

Early extrusion, *P. barbadensis* female 1.

molt without her, but they may have been dead and the gestation temperature they were kept at was not reported. Predation on a female's own eggs is unusual, since *D. diadema* and *P. marginemaculatus* females that drop infertile eggs or dead embryos will not even attempt to eat the broodsac no matter how hungry they are. (Failed sacs are always found at the bottom of the cage.) This female *P. barbadensis* was not starving.

Captive-hatched *P. barbadensis* 2nd instars had been offered by at least one vendor, though it is possible the number of successful hatches might be counted on a single finger. At least the one imported specimen was still doing well after three years, but this species was probably not imported again until 2012. In various past shipments, specimens were collected and packed well. Import DOA rate has been below 5% and most arrive with all appendages. There are two problems with acquiring this species. First, there is little demand for a species that is thought of as only slightly larger than our native Florida species.

Second, the Barbados export list is very limited in diversity and value. This makes it difficult for an importer to cover expensive shipping and paperwork costs, since the list only consists of a few crab species, one amblypygid, and one centipede species, none of which are in high demand (Gearheart pers. comm.).

Broods: 30-40.

Distribution: Barbados, St. Vincent, and the Grenadines.

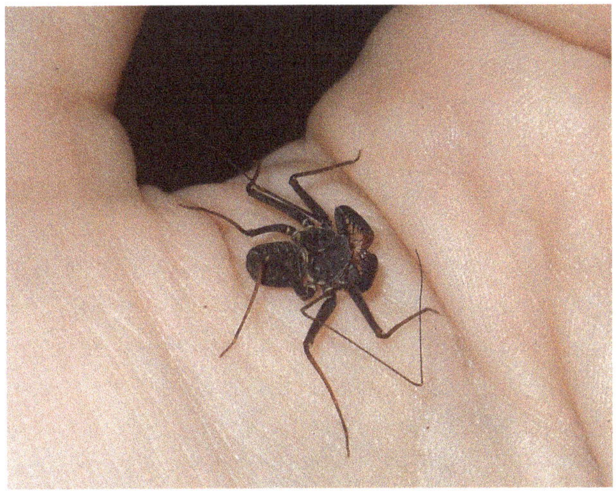

Immature *P. marginemaculatus* like most species are still quite tiny when they first obtain adult coloration (not sexually mature).

Phrynus marginemaculatus Koch 1841
SIDESPOTTED WHIPSPIDER

From 1993 to 2003 this was the only species that could be regularly acquired any time of the year. Interest in this species seemed to have dissipated when the much larger *D. diadema* became readily available. (The cost for this relatively tiny animal is nearly the same as *Damon.*) Specimens on vendor lists all but disappeared for a few years. Although specimens were difficult to acquire during the heyday of *D. diadema* imports, they have been consistently available again since around 2009.

Measurement of body size, *P. marginemaculatus.*

The sidespotted whipspider is found across peninsular Florida as well as a number of Caribbean islands. It is found inland and near shore, and is especially adapted to an intertidal habitat. It is reported to survive a full day submerged in seawater (Hebets and Chapman 2000). As far as I know, this is a very rare amblypygid ability, but some endemic Amazonian whipspiders may be adapted to short-term submersion as a result of extensive seasonal flooding of the habitat (though I cannot bring myself to throw an *H. batesii* in a fish tank). Room temperature works well in captivity, but Florida specimens, at least, can survive extremely cold temperatures for short periods as described in habitat temperatures.

Unfortunately not only does this magnificent arachnid order have to borrow common names from other well-known orders, but it has even lost at least one generic designation due to its relative obscurity. The sidespotted whipspider is often listed as *Tarantula marginemaculata*, but the generic term has been used for the popular group of pet Mygalomorph spiders, sometimes for Lycosid wolf spiders, and in reference to Tarantula hawks, etc. To avoid confusion with the vernacular, the real genus of this and related species has been suppressed by petition (Quintero 1981) and the next oldest generic description is used (*Phrynus*). In order to match the gender of the new genus, it is called *P. marginemaculatus* instead of *P. marginemaculata*. The generic redesignation is fine for tarantula hobbyists, but it certainly makes the amblypygid family Phrynidae formerly known as Tarantulidae more difficult to keep straight from the similarly spelled Phrynichidae.

Sexual maturity is reached at approximately 0.5" (12 mm) in body length, but adults can grow up to 0.7" (18 mm). However, Caribbean specimens must grow a little larger as even large, older Florida specimens only grow to 0.6" (16 mm)—though even at this size the whipspan can reach 4.5" (114 mm). As the name suggests, mature specimens are usually dark brown, almost black, with light spots (maculations) on the side margins of the prosoma. (However, similar markings are found on immature *P. barbadensis*.) 2nd instars have orange pedipalps, dappled markings, and a green abdomen, but this color gradually changes over to adult color with each molt.

Males and females display some slight differences, but are extremely difficult to tell apart by comparing the relative size of the palps and abdomen. On average, males are a little smaller and have a thinner abdomen, but the difference is slight and inconsistent. Older males also have larger palps relative to the body, but the difference is hard to quantify. Adults of this species are not the easiest to sex, though the high ratio of wild-caught females can make it seem more difficult than

Phrynus marginemaculatus spermatophore.

it is. The structure of the genital operculum is only slightly different at maturity, whereas in *Damon* and the closely related *Phrynus barbadensis* the greatly differing structure makes sexing rapid and definite. A check of the shape and size of the genital operculum on the underside of the abdomen may provide the best way to determine gender, as it is rounded towards the end and slightly larger in males. If a group of specimens are adults and the operculums all look exactly the same, they are probably all females. Differences in the genital operculum are not visible on immatures.

Courtship involves little to no physical contact other than with the whips. In communal caging the pair clears out a prime mating surface and keep cage mates away for a few days to a week. The male strokes the female with his antenniform legs, which results in hours of back and forth touching if she remains receptive (Weygoldt 1968). Only in the presence of a sexually mature female does the male glue a spermatophore to a vertical surface, which she can take up under her genital operculum. He may try mating again a few days later, but she can only take up one

Group of *Phrynus marginemaculatus*.

spermatophore—extras appear to be left untouched and then eaten the following night. The white spermatophore is 0.16″ (4 mm) and has a narrow stalk with widened structures at the apex as pictured.

A few months after mating the female extrudes the eggs inside a gelatinous, clear sac that hardens over the following days. Broodsacs are not produced by unmated females. The pastel green color of the eggs is visible on the first day or two, but the egg sac membrane darkens and the eggs appear brown for the rest of the gestation. Unlike the popular *Damon diadema*, the broodsac does not have a flat bottom and the side view is not blocked entirely by pleural folds.

Ova hatch after four months into white and green 1st instars that climb onto the mother's back and cling tightly. After nine or ten days they molt into 2nd instars that are ready to disperse and will soon hunt. Immatures can be kept together for a number of molts and may be reared with the mother. The primary concern for the early stages of this species is desiccation, because they are tiny and dry out rapidly.

Similar sized individuals can be kept in community cages. A 2.5-gallon (9 L) terrarium is plenty of space for half a dozen adults. This species seems to be especially tolerant of cohabitation, but this may be as much because adequate housing is easier to provide as because males are hardly aggressive towards each other. One or two inches of damp substrate is needed to help maintain humidity if ambient humidity is not high. Pieces of flat cork bark or wood leaned up against the sides or other pieces should be provided for necessary shelter and molting surfaces. In nature it may be found in shelters under old trees and stumps. Specimens have been found taking shelter together (Floyd pers. comm.).

Small specimens hide constantly if kept with larger animals. Unlike other species listed here, small specimens especially, display thigmotactic tendencies and are often found under bark or wood with just a narrow gap above the substrate. Molting surfaces in addition to hides should be made available since they hide in narrow spaces but will need to hang for molting. Dozens

Notice the spines on the front of the prosoma. *Phrynus marginemaculatus*.

can be kept together in relatively small caging, but problems may eventually arise from stress and territorial disputes. Cannibalism has only been reported in association with molts and is reduced if specimens are the same size, well-fed, and are given sufficient space. An adult will sometimes eat part of another that has recently molted, which invariably leads to death. It is hard to say if such deaths result from the keeper disturbing the recently molted animal, which then causes it to flee and invade another's territory, or if cannibalism is an accidental artifact of unnatural proximity. If a freshly molted animal is noticed, it should be placed in another cage for a week to allow the exoskeleton to harden. Sidespotted whipspiders are easy to house and more accepting of each other than most amblypygids. Keep in mind that despite gregarious tendencies, molting animals may still be killed.

Second instars reach adulthood in under two years with adequate feeding. Adults continue to molt after reaching maturity and can gain measurable size at this time. Sexually mature specimens usually live two to three years in captivity, for a total lifespan that is commonly less than five years.

Collection Report, courtesy Paul Bollinger: "In the past five years I have collected over a thousand adults. Populations are typically female heavy by a three to one ratio (also, I have been surprised at how small some of the gravid specimens are.) I currently live in Lee County which is situated on the southwestern coast of peninsular Florida. I have collected this species year-round to varying degrees of success and in various habitats. I have only observed them in my county and in the Keys.

"In the Keys, prominent areas consist of coral base rock that has very little organic decomposition on the surface. This habitat is usually void of whipspiders. One location with a sizeable population in the upper keys is situated among primarily gumbo limbo trees. It is an old dumpsite of coral limestone rocks. What makes this spot unique is a fair amount of organic decomposition. There are many old leaves among the coral rock and it is also heavily shaded. These factors are important because prey items are dependent on its organic decomposition. In the Keys location, the *Phrynus* are mainly under the coral stone.

"In Lee County the situation is different. Coastal southwestern Florida has much more organic topsoil. Since few areas have natural limestone exposed, the *P. marginemaculatus* here are almost exclusively found under pine and oak logs as well as logs of an invasive species known as the paper tree, *Melaleuca quinquenervia*.

The old pump house where numerous *P. marginemaculatus* are still found has grown closer to ground level over the last few years, Feb. 2013. © Paul Bollinger

"Cockroaches are a common factor in every population I have seen. They offer food to *Phrynus* at all phases of their life cycle. Cockroaches are not an exclusive prey item, but likely are the primary prey for all populations I have observed.

"The most specimens harvested in a day from one 'colony' was 54 adults when I first located this population. The site is an old, dilapidated pump house that contains rotting pine lumber and roofing shingles. It is approximately 3 ft (1 m) tall and 15 ft (3 m) in diameter. This site sits off the road about 50 ft while the surrounding trees include pine, palmetto, paper tree, and wax myrtle. I collect twenty or more specimens from this site three to four times a year and it continues to produce these numbers.

"In natural situations where no man-made trash structures exist, I primarily observe them under and inside rotting logs. The largest number of adults I have observed together was during a dry spell. They were under a palm log and there were ten adults and a few babies."

Broods: Females only produce 10-18 young at a time. Higher numbers may be possible for older, larger specimens found in the West Indies.

Distribution: Florida (USA), Bahamas, Cuba, Dominican Republic, Haiti, and Jamaica (Quintero 1981).

Phrynus operculatus Pocock 1902
RED BROWN WHIPSPIDER

Body coloration is very unusual for a whipspider since it is reddish brown to orange. This relatively colorful species is a bit larger than *Phrynus marginemaculatus* and has a more elongate body shape and extremities. Likewise, the scientific name is a little storied as Herbst (1797) placed it in the genus *Phalangium* following the lead of Linnaeus, but it could not remain there because that is a genus of Opiliones. The species name used by Herbst (1797) was *palmatum* but was suppressed by petition (Quintero 1981) because it was out of use for a time and used incorrectly for specimens of many other *Phrynus* (*Tarantula*). The species name *operculatus* refers to the abnormally enlarged genital operculum of the adult male. The only specimen I was ever able to acquire was collected from rock faces south of Big Bend in Texas in 1996.

It survived for eighteen months. This specimen did well with medium to high humidity and had a healthy appetite for small crickets. Quintero (1981) considered this a strictly surface dwelling species that lives close to the ground and explained specimens have been found under bark in at least three different Mexican localities as well as under rocks and stones.

In the Wild (Davenport pers. comm.): "On Big Springs Ranch, Texas, the animals come inside the buildings when the first cold weather hits from mid-October to early November. The habitat around the area where they are located is limestone (Cretaceous) with numerous small caves and crevices, oak and juniper mixed woodlands, and clear limestone gravel streams. The biotic zone where they are located is known as the Edwards Plateau. Other large arthropods in close association include the very large *Scolopendra heros castaneiceps* centipedes (large greenish black body with red head and yellow legs), small black scorpions, large stick insects, katydids, and yellow bark scorpions. They move quickly when approached so catching one is extremely difficult. When we see them, they are in the halls and on the walls inside the buildings and on rock cliffs and ledges outside, even climbing the limestone block walls of the buildings.

Broods: Possibly similar to *P. marginemaculatus*, but Quintero (1981) states a female from Mexico was found with 42 immatures.

Distribution: In Texas this species occurs over hundreds of miles that range across the Edwards Plateau, west through the Big Bend area and northwest almost into New Mexico. Texas (USA), northern Mexico, Guatemala (?) and Venezuela (?).

Phrynus operculatus, Big Springs Ranch, Real County, Texas. © Tripp Davenport

Phrynus operculatus, 1996.

Phrynus whitei with 1st instars. © Michael Seiter

Phrynus whitei Gervais 1842
HEADLIGHT WHIPSPIDER
This species is found widely across Central America and can vary greatly in size by geographic location. The stock that had been kept in the USA was double the size of *P. marginemaculatus* (similar to *P. barbadensis*), but did not reach the maximum 29 mm (1.14") body length (Viquez 2003) listed for this species. Although variable in many features, *P. whitei* has two very noticeable light-colored markings or "headlights" on the front of the prosoma, between the central and lateral eye clusters. While large males can have relatively larger pedipalps, they do not have notably exaggerated pedipalps like *Damon* or *Heterophrynus*.

Broods: 15-40.

Distribution: Phrynus whitei is found in Costa Rica, El Salvador, Guatemala, Honduras, Mexico, and Nicaragua (Viquez 2003).

GLOSSARY AND ANATOMY

ADULT STAGE Any time after the molt when a whipspider first becomes developed enough to be capable of mating or reproducing: first stage adult or stage 1 is initial sexual maturity, stage 2 is after an adult has molted once, stage 3 after it molts again, and so on. In this text stages are used rather than adult instar because instar is traditionally reserved for immature stages of arthropods and use for adults would blur an already imperfect line relative to sexual maturity in arthropods that continue to molt after maturity.

ANTENNIFORM LEGS Arachnid legs used in a wholly sensory function that do not contact the ground during walking and lack visible claws.

APOPHYSIS Fixed, blunt extension of the palp.

AUTOTOMY Reflexive release of an appendage at a specific point of attachment in order to facilitate escape from a predator or to regenerate a damaged appendage.

BASITARSUS Basal or proximal segment of the tarsus.

BOOK LUNG An air-filled space containing a respiratory organ filled with hemolymph (blood) that resembles a layer of flat sheets; located ventrally within the opisthosoma.

BROODSAC Extruded membranous sack containing a single clutch of eggs and attached to the female's opisthosoma; principally different from the egg sacks of insects and most arachnids because the eggs are never exposed to the atmosphere.

CAPTIVE BRED (C.B.) Offspring produced by adults reared to maturity in captivity.

CAPTIVE HATCHED (C.H.) Offspring that result from wild-collected adult females, even if mated in captivity.

CEPHALOTHORAX Prosoma; fused head and thorax common to arachnids and crustaceans.

CHELICERA (*pl.* CHELICERAE) Paired, pincer-like appendage in front of the mouth, unique to arachnids and other members of the subphylum Chelicerata.

COXA (*pl.* COXAE) Basal leg segment or hip.

CREPUSCULAR Active primarily at dusk and dawn.

DEIMATIC DISPLAY The defensive posture of a whipspider that includes reared-up palps and may include sound; threat display.

DENTICLES Raised bumps or small, blunt tipped spines that are part of the surface of the exoskeleton.

DISTITARSUS The terminal or distal segment of the palp or leg tarsus.

DISTAL Surface or segment furthest from the point of attachment.

DORSAL The upper surface or top.

INSTAR Developmental stage between each molt and prior to maturity.

ENTOMOPHAGUS MOLD Fungus that feeds on arthropods.

EXOSKELETON The external body covering of arthropods that serves as protection and muscular support. Mainly consisting of chitin and proteins.

EXUVIUM (*pl.* EXUVIAE) The shed exoskeleton or molt. (Exuvium is common usage and makes it easy to differentiate singular and plural, whereas "Exuvia" could be read as plural of exuvium or singular of exuviae.)

F1: Offspring of wild-caught adults (the offspring of two F1 would be F2 and so on). Wild-caught would be the P or parental generation so the sequence would be P, F1, F2, F3, etc. The F stands for filial and does not necessarily relate to wild-caught versus captive-bred, but use for most wild invertebrates in field observations or mixed source breeding (such as an F2 X F5 to produce an F6 or F3) would be highly dubious.

FEMUR (*pl.* FEMORA) Thigh; segment of an insect leg between the trochanter and tibia.

FLAGELLUM Thin, threadlike appendage, normally without its own muscles, but moved by muscles near the base of attachment; found on whipscorpions.

GENITAL OPERCULUM Large sternite containing the first pair of book lungs and gonopods which shields genitalia on the following segment. This is the first segment at the base of the opisthosoma after the pedicel.

GONOPODS Secondary sexual organs.

HAEMOCOEL Invertebrate body cavity through which hemolymph (blood) flows.

INTERSEGMENTAL MEMBRANE Elastic, less sclerotized portion of the exoskeleton joining segments.

LATERAL On the sides.

MARGIN A border or edge.

MEDIAN Central or middle; the central eyes in arachnids are known as median eyes.

MEDIUM (*pl.* MEDIA) The food as well as substrate used in culturing fruit flies, bacteria, etc.

Acanthophrynus in Jalisco, Mexico. Palpal tarsi fold back against the chelicerae. There are no pincers that fold up closed as in pseudoscorpions and scorpions. © Eddy Hijmensen

MOLT The act of shedding the exoskeleton.

MUSHROOM BODY Area of the arthropod brain associated with learning and spatial memory.

OCELLUS (*pl.* OCELLI) Light detection organs.

OPISTHOSOMA Last tagma of the arachnid body; contains the book lungs and reproductive organs.

PARTHENOGENESIS Reproduction in which unfertilized eggs develop into young. This is an exceptional reproduction method in the Amblypygi outside of one or two of the tiny *Charinus* spp.

PATELLA (*pl.* PATELLAE) Arachnid leg segment between the femur and tibia (in humans this is the small bone known as the knee cap).

PEDICEL Narrow waist (narrow first segment of the opisthosoma) connecting the prosoma and opisthosoma in certain arachnid orders including the spiders, whipspiders, and whipscorpions.

PEDIPALPS Palps; pair of large, leg-like mouthparts located outside the chelicerae; modified into a folded trap mechanism terminating in hand-like pincers in amblypygids.

PLEURAL MEMBRANE A stretchable, lightly sclerotized band of the exoskeleton which joins the abdominal tergites and sternites.

POSTEMBRYO An early, immobile developmental stage in arachnids that occurs between egg and first instar.

POSTULTIMATE MOLT A molt that takes place after sexual maturity has been reached and after the ultimate molt.

PROSOMA Cephalothorax; front tagma of the arachnid body; fused head and thorax from which the legs, eyes, and mouthparts originate.

PULVILLUS (*pl.* PULVILLI) Adhesive, fleshy pad located between the tarsal claws which allows an amblypygid to climb smooth surfaces such as glass. In whipspiders this is found only in the families Charinidae and Charontidae.

SCLEROTIZE To cause rigidity of the exoskeleton through addition of the protein sclerotin or substance other than chitin.

SETA (*pl.* SETAE) Hollow growth from the epidermal layer of the body wall; hair-like setae are often referred to as "hair" but are principally different from the hair of mammals.

SEXUALLY DIMORPHIC Presence of noticeably different external gender characteristics.

SPERMATHECAE Female's paired, sperm storage compartments.

SPERMATOPHORE Structure formed by the male which includes sperm and is used in the transfer of sperm to the female.

STERNITE Ventral abdominal plate or segment.

STRIDULATE To make sound by rubbing together specialized body parts. A few genera of whipspiders have this ability.

TAGMA (*pl.* TAGMATA) Grouping of multiple segments forming a functional group, e.g. thorax, cephalothorax, opisthosoma, and prosoma.

TARSUS (*pl.* TARSI) The foot of an arthropod. The tarsus contains a number of small segments including the claws.

TENERAL Soft, often pale or white colored state of the exoskeleton following a molt and before it has mostly hardened or darkened (the hardening process is call sclerotization or tanning).

TELSON Terminal segment of the scorpion's abdomen.

TERGITE Dorsal abdominal segment.

THIGMOTACTIC A preference to feel surfaces against the top and bottom of the body.

TIBIA (*pl.* TIBIAE) Segment of an arachnid leg between the tarsus and patella.

TRICHOBOTHRIA Sensitive, long setae located in membranous socketed pits and commonly found in specific positions on the palps and legs, used primarily to aid in prey capture by detecting movement and vibration. The arrangement can be an important taxonomic character since they are relatively invariable.

ULTIMATE MOLT The final shedding of the exoskeleton from an immature and resulting in a sexually mature specimen.

BIBLIOGRAPHY

Alexander, A. J. (1962) Biology and behavior of *Damon variegatus* Perty of South Africa and *Admetus barbadensis* Pocock of Trinidad, W.I. (Arachnida, Pedipalpi). *Zoologica* 47: 25-37.

Armas, Luis F. de. (2005) Notas sobre la biologia reproductiva del amblipigido partenogenetico *Charinus acosta* (Quintero, 1983) (Amblypygi: Charinidae). *Boletin. Soc. Ent. Aragonesa* 36: 271-273.

Beccaloni, Jan. (2009) *Arachnids*. University of California Press, Berkeley and Los Angeles, CA.

Beck, L. (1968) Aus den Regenwaldern am Amazonas II. *Natur and Museum* 98: 71-80.

Beck, L., and K. Görke. (1974) Tagesperiodik, Revierverhalten und Beutefang der Geißelspinne *Admetus pumilio*. C. L. Koch im Freiland. *Z. Tierpsychol.* 35: 173-186

Breene, Robert G., III (1995) Tarantulas finally catch up. *Forum of the American Tarantulas Society*. 4(1): 24-26.

Cernicka, Martin. (2003) *Damon* sp. (Tanzania). *Sklipkan Journal for Invertebrates Keepers and Breeders* 2/2003: 52.

Dias, Sidclay C., and Glauco Machado. (2006) Microhabitat use by the whip spider *Heterophrynus longicornis* (Amblypygi, Phrynidae) in Central Amazon. *The Journal of Arachnology* 34: 540-544.

Ditzel, Nicholas. (2002) Keeping and breeding amblypygids (whip spiders). *Invertebrate* (The Society Invertebrate, Denmark) 4(Fall): 28-34.

Dixey, Steven. (2002) The Tanzanian giant banded tailless whipscorpion (*Damon diadema*). *Invertebrates-Magazine* (June): 10+.

Dunlop, Jason A. (1995) Gigantism in arthropods. *Forum of the American Tarantula Society*. 4(5): 145-147.

Evans, Arthur V. (2007) *Field Guide to Insects and Spiders of North America*. Sterling Publishing Co., New York, NY.

Foelix, R. F., I. Chu-Wang, and L. Beck. (1975). Fine structure of tarsal sensory organs in the whip spider *Admetus pumilio* (Amblypygi, Arachnida). *Tissue and Cell* 7(2): 331-346.

Foelix, R. F., and D. Troyer. (1980) Giant neurons and associated synapses in the peripheral nervous system of whip spiders. *J. Neurocytol.* 9: 517-535.

Foelix, R. F., and E. A. Hebets (2001) Sensory biology of whip spiders (Arachnida, Amblypygi). *Andrias* 15: 129-140.

Foelix, R., D. Troyer, et al. (2002) Peripheral synapses and giant neurons in whip spiders. *Microscopy Research and Technique* 58: 272-282.

Forster, R. R., and L. M. Forster. (1970) *Small Land Animals of New Zealand*. Revised Edition. John McIndoe Limited, Dunedin.

Frye, Frederic L. (1992) *Captive Invertebrates: A Guide to Their Biology and Husbandry*. Krieger Publishing Company, Malabar, FL.

Giupponi, A. P. de L. (2002) Duas novas sinonímias no genero *Heterophrynus* Pocock, 1894 (Amblypygi: Phrynnidae). *Revista Iberica de Aracnologia* 6(31): 173-175.

Harrell, Wade. (2006) Tailless whipscorpions: the Amblypygi. *The ATS Forum Magazine*. 15(4): 9-14.

Harvey, Mark S. (2002) The first Old World species of Phrynidae (Amblypygi): *Phrynus exsul* from Indonesia. *The Journal of Arachnology* 30: 470-474.

Harvey, Mark S. (2003) *Catalogue of the Smaller Arachnid Orders of the World: Amblypygi, Uropygi, Schizomida, Palpigradi, Ricinulei and Solifugae*. CSIRO Publishing, Collingwood, Victoria, Australia.

Harvey, Mark. (2011) Order Pseudoscorpiones de Geer, 1778. *In*: Zhang, Z.-Q. (Ed.) Animal biodiversity: An outline of higher-level classification and survey of taxonomic richness. *Zootaxa* 3148: 120.

Harvey, Mark S., and Paul L. J. West. (1998) New species of *Charon* (Amblypygi, Charontidae) from Northern Australia and Christmas Island. *Journal of Arachnology* 26: 273-284.

Hebets, Eileen A. (2002) Relating the unique sensory system of amblypygids to the ecology and behavior of *Phrynus parvulus* from Costa Rica (Arachnida, Amblypygi). *Canadian Journal of Zoology* 80: 286-295.

Hebets, E. A., and R. F. Chapman. (2000) Surviving the flood: Plastron respiration in the non-tracheate arthropod *Phrynus marginemaculatus* (Amblypygi; Arachnida). *Journal of Insect Physiology* 46: 13-19.

Hillyard, Paul. (1995) *Spiders and Scorpions*. Reader's Digest Young Families.

Kirby, Virginia. (1986) Amblypygids. *Backyard BugWatching* No. 1 (Fall)

Immature *Paraphrynus raptator* with large cricket. © Paul Bollinger

Kury, A. B., and R. Pinto-da Rocha. (2002) Opiliones. In: Adis, Joachim, Editor (2002) *Amazonia Arachnids and Myriapoda*. Pensoft Publishers Sofia, Bulgaria.

Ladle, Richard J., and Kathryn Velander. (2003) Short communication: fishing behavior in a giant whip spider. *The Journal of Arachnology* 31: 154-156.

Levi, Herbert W., and R. Lorna. (1968) *Spiders and Their Kin*. Western Publishing Company, Racine, WI.

McMonigle, Orin. (2005) Enter the *Damon*: captive reproduction. *Invertebrates-Magazine* 4(4): 6-10.

McMonigle, Orin. (2006) *Giant Tarantulas: The Enthusiast's Handbook*. Elytra and Antenna, Brunswick, OH.

McMonigle, Orin. (2008) *Whipscorpions and Whipspiders: Culturing Gentle Monsters*. Elytra and Antenna, Brunswick, OH.

McMonigle, Orin. (2009a) *Scorpions, Windscorpions, Pseudoscorpions: Culturing Ancient Arachnids*. Elytra and Antenna, Brunswick, OH.

McMonigle, Orin. (2009b) Sidespotted whipspiders: *Phrynus marginemaculatus*. *Invertebrates-Magazine* 8(4): 4-8.

McMonigle, Orin. (2011) The latest word in whipspider species' imports, part 1. *Invertebrates-Magazine* 11(1): 5-8.

McMonigle, Orin. (2012a) The latest word in whipspider species' imports, part 2. *Invertebrates-Magazine* 11(2): 2-6.

McMonigle, Orin. (2012b) The captive reproduction and life cycle of the cosmetid *Vonones ornata*. *Invertebrates-Magazine* 11(3): 5-8.

McMonigle, Orin. (2013a) *Keeping the Praying Mantis: Mantodean Captive Biology, Reproduction, and Husbandry*. Coachwhip Publications, Greenville, Ohio.

McMonigle, Orin. (2013b) A prehensile "tail" in the Amblypygi. *Invertebrates-Magazine* 12(2): 16,18.

Mullinex, C. (1979) A new *Paraphrynus* from Yucatan (Amblypygida, Tarantulidae). *Journal of Arachnology* 7(3): 267-269.

Preston-Mafham, Rod, and Ken Preston-Mafham. (2005) *Encyclopedia of Insects and Spiders*. Thunder Bay Press, San Diego, CA.

Prendini, Lorenzo. (2011a) Order Schizomida Petrunkevitch, 1945. *In*: Zhang, Z.-Q. (Ed.) Animal biodiversity: An outline of higher-level classification and survey of taxonomic richness. *Zootaxa* 3148: 156.

Prendini, Lorenzo. (2011b) Order Solifugae Sundevall, 1833. *In*: Zhang, Z.-Q. (Ed.) Animal biodiversity: An outline of higher-level classification and survey of taxonomic richness. *Zootaxa* 3148: 118.

Prendini, Lorenzo. (2011c) Order Amblypygi Thorell, 1883. *In*: Zhang, Z.-Q. (Ed.) Animal biodiversity: An outline of higher-level classification and survey of taxonomic richness. *Zootaxa* 3148: 154.

Prendini, Lorenzo. (2011d) Order Thelyphonida Latreille, 1804. *In*: Zhang, Z.-Q. (Ed.) Animal biodiversity: An outline of higher-level classification and survey of taxonomic richness. *Zootaxa* 3148: 155.

Punzo, Fred. (1998) *The Biology of Camel-Spiders (Arachnida, Solifugae)*. Kluwer Academic Publishers, Norwell, MA.

Quintero, D. (1979) Sobre *Paraphrynus emaciatus* Mullinex, *P. leptus* Mullinex (Amblypygi: Phrynidae) y el dimorfismo sexual en los amblypygidos. *Cuaderno de Ciencias* 3. pp. 15-24.

Quintero, Diomedes, Jr. (1981) The amblypygid genus *Phrynus* in the Americas (Amblypygi, Phrynidae). *The Journal of Arachnology* 9: 117-166.

Rayor, Linda S. (2007) Family ties: unexpected social behavior in an improbable arachnid, the whip spider. *Natural History* (Feb.): 38-44.

Rayor, Linda S., and Lisa A. Taylor. (2006) Social behavior in amblypygids, and a reassessment of arachnid social patterns. *The Journal of Arachnology* 34: 399-421.

Romero, Frances. (2008) Top 10 animal stories of 2008, 7. World's longest insect. *Time* (Nov. 3).

Santer, Roger D., and Eileen Hebets. (2009) Tactile learning by the whip spider, *Phrynus marginemaculatus* C. L. Koch (Arachnida, Amblypygi). *Journal of Comparative Physiology A* 195(4): 393-399.

Santer, Roger D., and Eileen A. Hebets. (2007) Agonistic signals received by an arthropod filiform hair allude to the prevalence of near-field sound communication. *Proceedings of the Royal Society* 275: 363-368.

Schleser, David M., and David T. Roberts. (1997) Extra-terrestrials: the fascinating world of tailless whipscorpions. *Fauna* 1(2): 46-52.

Sejna, Vladimir. (2005) Schizomida. *Sklipkan Journal for Invertebrates Keepers and Breeders* 1/2005: 32.

Shear, William A. (1970) Stridulation in *Acanthophrynus coronatus* (Butler) (Amblypygi, Tarantulidae). *Psyche* 77: 181-183.

Spence, A. J., and E. A. Hebets. (2006) Anatomy and physiology of giant neurons in the antenniform leg of the amblypygid *Phrynus marginemaculatus*. *The Journal of Arachnology* 34(3): 566-577.

Stanek, V. J. (1969) *The Pictorial Encyclopedia of Insects*. Hamlyn House, London.

Stoops, Erik D., and Jeffrey Martin. (1995) *Scorpions and Venomous Insects of the Southwest*. Golden West Publishers, Phoenix, AZ.

Strausfeld, N. J., Hansen, L., Li, Y., Gomez, R. S., and K. Ito. (1998) Evolution, discovery, and interpretations of arthropod mushroom bodies. *Learn. Mem.* 5: 11-37.

Viquez, Carlos. (2003) The whip spiders from Costa Rica (Amblypygi, Arachida). *Sklipkan Journal for Invertebrates Keepers and Breeders* 4/2003: 124.

Walsh, Rachel E., and Linda S. Rayor. (2008) Kin discrimination in the amblypygid, *Damon diadema*. *The Journal of Arachnology* 36: 336-343.

Werner, F., and C. Olson. (1994) *Insects of the Southwest*. Fisher Books, Tucson, AZ.

Weygoldt, Peter. (1968) Beobachtungen zur fortpflanzungsbiologie und zum verhalten der geißelspinne *Tarantula marginemaculata* C. L. Koch (Chelicerata, Amblypygi). *Zeitschrift zur Morphologie der Tiere* 64: 338-360.

Weygoldt, Peter. (1969) *The Biology of Pseudoscorpions*. Harvard Books in Biology, Number 6. Harvard University Press, Cambridge, MA.

Weygoldt, Peter. (1977) Coexistence of two species of whip spiders (Genus *Heterophrynus*) in the neotropical rainforest (Arachnida, Amblypygi). *Oecologia* (Berl.) 27: 363-370.

Weygoldt, Peter. (1999) Spermatophores and the evolution of female genitalia in the whip spiders (Chelicerata, Amblypygi). *The Journal of Arachnology* 27: 103-116.

Weygoldt, Peter. (2000) *Whip Spiders (Chelicerata: Amblypygi): Their Biology, Morphology and Systematics*. Apollo Books, Denmark.

Weygoldt, Peter (2002): Amblypygi. In: Adis, Joachim, Editor (2002) *Amazonia Arachnids and Myriapoda*. Pensoft Publishers Sofia, Bulgaria.

Weygoldt, Peter. (2003) Reproductive biology of two species of *Phrynichus, P. exophthalmus* Whittick, 1940 and *P. deflersi arabicus* nov. ssp. (Chelicerata: Amblypygi). *Zoologischer Anzeiger* 242(3): 193-208.

Coachwhip Publications

CoachwhipBooks.com

THE ULTIMATE GUIDE TO BREEDING BEETLES
Orin McMonigle
ISBN 1-61646-132-2

KEEPING THE PRAYING MANTIS
Orin McMonigle
ISBN 1-61646-166-7

MILLIPEDS IN CAPTIVITY
Orin McMonigle
ISBN 1-61646-143-8

INVERTEBRATES FOR EXHIBITION
Orin McMonigle
ISBN 1-61646-105-5

www.ingramcontent.com/pod-product-compliance
Ingram Content Group UK Ltd.
Pitfield, Milton Keynes, MK11 3LW, UK
UKHW051548161225
9597UKWH00010B/110

9 781616 461836